Community Forest Management in Protected Areas

Van Gujjars Proposal for the Rajaji Area

Foreword
Justice P.N. Bhagwati

Rural Litigation & Entitlement Kendra

Natraj Publishers
Dehra Dun

First Published 1997

© *Rural Litigation & Entitlement Kendra, Dehra Dun.*

ISBN 81-85019-64-9

Published by Mrs. Veena Arora for Natraj Publishers, Publications Division, Dehra Dun and printed at Jaysons Quadra Colour Printers, New Delhi.

Table of Contents

(e) Increased Rate of Soil Erosion
(f) Forest Fires
(g) Conclusions: Ecological Problems and Human
Impact

Part III : Local People and Problems

Part IV. Community Forest Management in Protected Areas

LIST OF MAPS

Justice P.N. Bhagwati
Former Chief Justice of India
and presently Vice Chairman,
United Nations Human Rights Commission

S-296, Greater Kailash-II,
New Delhi-110048

FOREWORD

I have read with great interest the plan prepared by the Van Gujjar community for forest management of the Rajaji area. I believe this is the first time that, in the preparation of a plan affecting the people, effective participation of the people themselves has been secured. Unfortunately, in our country, the governments as well as non-governmental organisations adopt a paternalistic attitude, and come out with plans for what they regard as beneficial to the disadvantaged and vulnerable sections of the community, without consulting the affected people or ensuring their participation in making of the plans. The result is that the plans, which the Governments and even non-governmental organisations make for relieving the poverty and distress of the have-nots and the handicapped, remain highly elitist documents not capable of effective implementation and, even if implemented, they do not help to alleviate the suffering of the people. I am, therefore, very glad that for the first time a plan has been prepared with the effective participation of the people who are going to be the beneficiaries of the plan.

I find that, unfortunately, laws for the protection of the forests are highly theoretical and unjust, and they do not take into account the socio-economic realities and needs of the people who are living in the forests and whose subsistence depends entirely upon the forest produce. These laws are conceived in the offices of the bureaucrats and the chambers of the concerned minister, regardless of the situation prevailing at ground level and totally ignoring the lifestyle of the forest denizens. The answer lies in including the people residing in the forest regions and eking out their daily living from forest produce, taking them into confidence and giving them an effective voice in the policy decisions affecting forest management. It is then that we can really have an effective plan for the daily requirements of the people and only then can we build up a movement at the grass root level for protection and preservation of the forests.

There is a common belief that tribals and other vulnerable sections of the community living in the forest regions are uncultured and uncivilised, since the benefits of the modern civilisation have not reached them. But this is a myth and an illusion; there is so much traditional wisdom amongst these people, accumulated over the years, that they are able to live in harmony with nature. Nature and environment are as much a part of their daily existence as food, shelter and clothing and they are continuously in communion with nature. Our plan for forest management must, therefore, take into account the human beings who live in the forests and nothing should be done which would affect their daily existence or their means of subsistence.

It has been rightly pointed out by the Rural Litigation & Entitlement Kendra that the Van Gujjars living in the forests of the proposed Rajaji National Park, as also other communities living in other similar areas, have been the victims of faulty environment conservation policies and because of such wrong policies, these communities are on the verge of extinction. It is, therefore, a matter of great satisfaction for me to learn that a plan for protected areas has been prepared which would take into account the symbiosis between the local communities and the forests in which they live and which would maintain social diversity as well as forest bio-diversity.

I congratulate RLEK for taking the initiative in the preparation of this plan. Its Chairman, Shri Avdhash Kaushal, is one of the most sincere and dedicated social activists I have met, and what he has done for the preservation of ecology and environment in the forest region of Dehra Dun and the surrounding areas is outstanding and deserves the highest praise. I hope and trust that he will continue to receive support and encouragement from all right thinking people, including the Government, in the stupendous task of ecological preservation, including forest management, in which he is engaged. However high the praise I may lavish upon him, it would not do justice to the wonderful work which he is doing in this region.

New Delhi
April 21, 1996

(P.N. Bhagwati)

PREFACE

Living in the forests of the proposed Rajaji National Park, the Van Gujjars, and other similar communities elsewhere, have been the victims of the faulty environment conservation policies. This caused the indigenous Van Gujjars community to reach the verge of extinction, eliminating yet another element of India's social and terrestrial diversity. It was imperative that an alternative forest management policy be devised, which would honour the symbiosis between such people and the forests they inhabit. The present plan for Protected Areas is the logical outcome. The unique feature about this plan is that it emanates from the understanding, aspirations, and confidence of the Van Gujjar themselves. The plan has been aptly named 'Community Forest Management in Protected Areas - Van Gujjars Proposal for the Rajaji Area'.

First and foremost, therefore, we acknowledge the traditional wisdom and the participation of the Van Gujjar people who were the main instruments in the making of this plan. Dr. Pernille Gooch of the Department of Anthropology, University of Lund, Sweden, was the first person to have done formal research among this community. It was her work that inspired Rural Litigation & Entitlement Kendra (RLEK) to intervene on behalf of the Van Gujjars in 1992. We acknowledge her inspired work showing her deep concern for community. It was during this that CEBEMO (now BILANCE), of the Netherlands, supported RLEK's intervention in organising the community, and we take this opportunity to thank them. Though the genesis of the ideas emanated from the community, its formalizing was team work by independent researchers. We acknowledge the immense effort put in by Dr. Alan Warner, Professor in the School of Education at Acadia University, Wolfville, Nova Scotia, Canada, who was the key actor in forming the plan, collecting and collating the primary data from the community, and writing the script. The Canadian IDA kindly paid Dr. Warner's travel and living expenses. Tremendous help and support was given by Dr. Edward Glanville, Professor in the Department of Anthropology, McMaster University, Ontario, Canada. The two Doctors were ably assisted by Mr. Simron Jit Singh, a Masters in Ecology and Environment from New Delhi and Miss Bindu Kalia, a Masters in Social work from Jamia Millia Islamia, New Delhi and our colleague at RLEK. We acknowledge their unstinted efforts in both field and desk work.

We are grateful for the contribution made by the members of the Student Study Group from the Law Faculty, University of Delhi, guided by their Dean, Faculty of Law, Prof. M.P. Singh and the Head of the Law Campus, Professor B.B. Pandey. The team spent considerable time and effort in interacting with the community and guiding the plan within the confines of the law of the land. We similarly acknowledge the immense help that Professor M.K. Ramesh of the Law School of India University, Bangalore, gave us in analysing the plan within the framework of the Constitution of India. Our grateful thanks to Ms. Nitya Ramakrishnan and Mr. Ashok Aggarwal, both practising lawyers of the Supreme Court of India, for their invaluable help in explaining the practice of law regarding a plan of this nature.

The Centre for science and Environment, New Delhi, has all through our effort supported RLEK on the issue of community forest management. We acknowledge the constant support given to this venture by Dr. Anil Aggarwal, Ms. Sunita Narain, Ms. Neena Singh and Mr. Ravi Sharma.

Acknowledgments are due to the Chief Conservator of Forests, UP, Mr. Shyam Lal (IFS), to the Conservator of Forests, Meerut Circle, Mr. Srikant Chandola (IFS), and to Mr. C.P. Goyal (IFS), Divisional Forest Officer (Social Forestry), Meerut Division, for the invaluable understanding of the people and forests from a Forest Department prespective. Acknowledgments are also due to the 62 participants of the workshop held in Feb. 96 who discussed and analysed the contents of the first draft. We also acknowledge the silent and almost invisible effort made by RLEK field workers who helped the research team in field visits and data collection.

The national print and electronic media have, of course, always supported RLEK interventions, particularly this one on the community forest management of protected areas. We wish to acknowledge the help of all those media persons who not only ran the stories but also actively participated in the workshop.

We are grateful to Mr. Justice P.N. Bhagwati, former Chief Justice of India, now Vice-Chairman of the United Nations Human Rights Commission, for writing the Foreword.

Last, but not the least, our grateful thanks go to the Swedish Society for Nature Conservation (Naturskydds Foreningen) whose concern and financial support made the plan possible, and to the International Workgroup for Indigenous Affairs (IWGIA), Denmark, for supporting the cause of indigenous people.

Avdhash Kaushal
Chairman
Rural Litigation & Entitlement Kendra
Dehra Dun

Introduction

There has been extensive policy discussion and debate in India over the past decade about the ecological viability of protected areas in relation to the needs of the local peoples who live in and around them (Down to Earth, 1995). Many different perspectives have been presented on the importance and problems of protected areas, with an increasing consensus, based on a wide range of evidence, that the international and national protected area policies of the 1970s and 1980s have failed both to protect ecosystems and the interests of the local people (Kemf, 1993; Kothari et. al., 1995).

A high profile example of this issue has been the conflict between the Van Gujjars, a forest-dwelling pastoralist tribe in Northern Uttar Pradesh, and the authorities responsible for the protected area in which they live, the proposed Rajaji National Park (see Map 1). The Rajaji - Van Gujjars case has attracted extensive media coverage (RLEK, 1995) and have become focal points in higher level policy discussions of the relationship between protected areas and local people (SPRIA & RLEK, 1993). Divergent positions have been articulated from the plan to remove the Van Gujjars forcibly (Verma, 1983), to the recommendation to move them "voluntarily" (Kumar, 1995), to the position that they may stay if they wish (Pottie, 1995), to the position that they manage the proposed park area (Gooch, 1994). Although there has been much writing and debate, it has been restricted to the general level of policy discussion lacking specific structures and proposals for reform and implementation. The only specific proposal was the original plan to evict the Van Gujjars by force to a new settlement built for them at Pathri. However, it is now widely recognised as a failure given the refusal of the Van Gujjars to re-locate under the original terms and the subsequent directive of the Uttar Pradesh government that their wishes be respected. Thus, despite extensive inquiry, writing and discussion, the debate has hardly moved beyond analysis and critique of the status quo and general proposals for the future. Yet through out the discussions, the Gujjars over the past several years have been asking for management of the forests e.g. the request of Mustooq Lambardar from the Gohri Range, "Give us the management and we will turn the forest into a diamond."

Within this context, a research team facilitated by Rural Litigation Entitlement Kendra (RLEK) began to work with the Van Gujjars to help them to propose a specific management model for resolving the problems in the Rajaji area. RLEK has been extensively involved in the Rajaji case as a strong advocate for the Van Gujjars, whilst it also has a long standing recorded commitment to working for environmental protection in the local area (RLEK, 1994). RLEK requested that this specific model be based on the expertise of the Van Gujjars and other local people while recognising the expertise of policy makers, activists, and researchers.

This report, based on seven months of field work and research, including many hours' of discussion in the forest with Van Gujjars and other local people, sets forth proposals which aim to provide:

- a practical model of community forest management in protected areas (CFM-PA) in which local people are the lead managers and the Forest Department role is that of supporter and monitor.
- a specific, practical set of structures through which to implement CFM-PA in the Rajaji area which assures environmental protection whilst respecting the needs, rights and traditions of local people.
- a means for furthering sound policy development and reform for protected areas at a national level.

The report has been divided into four major sections in order to achieve these objectives :

I. *Protected Areas and Indigenous People:* An analysis of past and present international and national policies for protected areas, demonstrating the need for legal reform and for alternative management models which protect both local ecology and needs of local people. This critique is followed by a review of the successes and limitations of alternative forest management practices in India and elsewhere and suggesting effective strategies to address the Rajaji problems and context.

II. *The Rajaji Ecological Context:* A description, analysis and critique of the local context and ecological problems in the Rajaji area, identifying the major contributing factors to problems, and the need for an alternative management model.

III. *Local Peoples Problems:* An analysis of the perspective of local people in relation to their traditional rights, use of forest resources, and conflict with park authorities. Specific attention is devoted to the Van Gujjars and the characteristics of their cultural and ecological values that would make them appropriate guardians of the forest.

IV. *Community Forest Management in the Rajaji Area:* A description of the plan procedure by which the Van Gujjars and other local people would manage the protected area facilitated and monitored by local authorities and agencies. These proposals are based on the research on effective practices in the local, national, and international context presented in first three parts.

As noted above, much has been written about past policies and practices, yet it is important to present these sections in detail as they form the framework for the specific community management alternative. Part four represents the major new contribution of this report; we particularly welcome discussion about these ideas. We encourage critics to take a constructive approach to this discussion by offering specific suggestions for overcoming obstacles to the improvement of these proposals. Inevitably, it is easier to criticise than to provide positive alternatives.

The problems between park authorities and local peoples in the Rajaji area are by no means restricted to issues between the park authorities and the Van Gujjars, as will be highlighted in the report. Although the Van Gujjars have received more media attention, the rights and needs of other groups such as Taungyas, border villages, and Gothiyas have also been ignored in protected area policies. The injustices to these groups is discussed in part three, with community management strategies proposed to rectify them in part four. We have devoted more attention to the role of the Van Gujjars because they are the only forest dwellers who inhabit the core areas of the proposed park. The Van Gujjars have been the primary targets of eviction and continue to be viewed as the worst problem by park authorities today, as protection of the core area has been a particular ecological priority. We are therefore proposing that the Van Gujjars become lead managers of the core area whilst other groups have lead management roles in relation to the resources and areas that they use.

Finally, it is important to analyse the failures of existing policies in order to justify the change to new systems. We believe the long history of failures is, in fact due to the Forest Department's lack of examining its own structures. In the past, problems and failures have either been blamed on outside groups or on misguided objectives, procedures, or individuals. The pattern of continual failure, despite new proposals over the years by higher level forest officials, is precisely because the structure of the system itself has produced the failures. We emphasise that it has been the failure of structures and systems rather than the failure of the people working within them. There are many good people working within the Rajaji context and documenting abuses; there is no intention to blame individuals or put others at risk.

Although we have provided detail on abuses, we have avoided data which could identify individuals. Even if particular individuals were replaced, the system would cause their replacements to continue past practices and the situation would not be improved. We must respect that the proposed Rajaji Park is a community of people and wildlife which continually must interact with each other, for better or worse. There are no "good" or "bad" individuals or groups, but rather a `dys' -functional system in which all groups participate, some with more power than others. Although this report directly criticises the forest management system, we believe the forest officials are as much victims of that system and subject to its abuses as are the wildlife, Van Gujjars, Taungyas, villagers, or Gothiyas. They have much to contribute to positive alternatives. The report aims to facilitate improvement of the system for the benefit of the environment and all the people living and working within it.

Description of the Rajaji Park

The proposed Rajaji Park comprises 825 square kilometres of land situated in the Shiwalik hills, and is representative of the Shiwalik ecosystem which lies between the Himalaya and the Upper Gangetic plains. The Rajaji area is among the better preserved remnants of the Shiwalik ecosystem and represents the Northwest limits of the tiger and elephant populations in India and includes a diversity of flora and fauna with a preponderance of sal forest.

Van-Gujjars are pastoral nomads and forest dwellers who earn their livelihood through raising buffaloes and selling milk. They utilise the forest for fodder for their animals. They have lived for centuries in deras, large thatched circular huts, throughout much of the park area as well as in the Shiwalik Forest Division to the west. The majority of Van Gujjars either migrate to the mountains for the summer and monsoon seasons, or move shorter distances out of the park to seasonal forest locales. There are also a large number of Taungya and local villagers living on the borders of the proposed park who have traditional rights to (and depend on) the forests for the collection of minor forest produce and grazing. A range of urban and government developments have encroached into the park area over the years, including the Chilla power canal and an army ammunitions camp which have blocked the major east-west elephant migration corridor.

Ecological Problems and the Impact of Human Beings

The report carefully analyses the status and contributing factors of six major ecological problems in the Rajaji area: forest deterioration, depletion of ground cover, weed infestation, wildlife problems, soil erosion, and forest fires. The official plan for the proposed park provides a range of information on these problems, but primarily attributes the negative impact to Van Gujjars, local peoples, and insufficient park resources. A detailed analysis of the official information, the findings provided by external empirical research studies, our field observations, and the perspectives of a wide range of Van Gujjars and villagers, invalidate this official park view. This report documents past destructive forest management practices, present poaching and tree felling, and the extensive system of illicit fees resulting in over exploitation by local peoples, as the principle causes of the ecological problems.

The ecological problems of the proposed Rajaji Park are due to overall structural problems in the management system. The system must be re-structured with decision-making authority placed in the hands of groups with the highest investment in forest protection. Behavioural incentives must be structured so as to encourage behaviours that reduce human impact on the environment. The management process must be open, monitored, and accountable, to deter any one group from exploiting the forest for its own benefit.

Local Peoples and Problems

Van Gujjar Problems & Perspectives: Van Gujjars see the major threats to their development and survival as (1) a fundamental insecurity and exploitation with respect to earning their livelihood, (2) poor education, and (3) poor health care. They recognise the widespread difficulties under the present forest management system. They view their culture and lifestyle as their greatest strength, and refuse to accept the notion that it is backward, impoverished, and should be abandoned. They are not prepared to give up their identity in return for the solutions offered by others, hence their continued refusal to move out of the forest. Their priority is to maintain this lifestyle which means addressing their demands for security in land and livelihood, access to education, and good health care. They want to control programmes and have them delivered in a form that supports their own culture and values. They have seen effective models for education and health programs in the forest. They have always believed that they would be far better managers of the forest than the Forest Department and cite endless examples to make their case. It is only recently however that they have begun to see that this is a realistic possibility given the continuing problems of the present system and the support they have received from outsiders. Community forest management has become the preferred option as they see that it could maintain their lifestyle, change the management system that has put it in jeopardy, and preserve the forest.

Villager Problems and Perspectives: The Taungya villagers argue that they gave up their livelihoods and/or property everything in their original villages to come and plant trees for the benefit of the Forest Department. After fifty years of service, they now find themselves with neither ownership of the land they have nurtured nor legal or traditional rights. Taungyas demand revenue village status, legal rights, and access to development programmes. They point out that their traditional occupation is to plant and grow trees, and that they have the expertise and interest in growing and protecting the forest, if they can be guaranteed access to the resulting minor forest products that they require for their subsistence.

Other villagers repeated the Taungya themes, in relating anger toward the Forest Department and its attempts to extinguish their traditional rights to the forest. Their foremost concern focuses on access to bhabbar grass which is the mainstay of their livelihood. They also want their traditional rights to grazing and firewood collection recognised. Villagers also complained about crop predation caused by animals living in the park area, particularly elephants. All the villagers we interviewed demanded that their traditional rights to the forest be recognised, and were all very supportive of the idea of organising forest protection committees through which they could manage the discrete forest areas that they regularly use on the borders of the park.

Culture and Ecological Values of the Van Gujjars
The Van Gujjars would seem to be particularly well fitted by their culture and past history to assume guardianship of the forest. Practical, realistic and down-to-earth, they possess intimate knowledge of wildlife and the forest. Cultural values and a feeling for the forest are still largely intact and these speak strongly against poaching or destruction of habitat. The values are enforced by sanctions acting in concert at levels ranging from social pressure and the need for self-respect, to customary law with an effective system of panchayats, to religion and supernatural sanctions. Their self-contained society

is a choice to live in their own universe of praise and blame, indifferent to the opinions of other people.

All forest animals are seen as interrelated components of the natural world and are therefore, to be cared for by mankind. No forest creature may be eaten or skinned and if found dead should be buried. Realistic in their approach to the forest, the Van Gujjars say pragmatically; "The forest is our home, why should we destroy it?" Thus, they take only what they themselves need and there is no tradition of selling forest products of any kind, whether wood, honey, rope or medicines. These ecologically based values have been resistant to change over many years, despite their constant trading relationships with the wider society, and presumably would be retained into the foreseeable future provided they are permitted to continue as a close-knit society.

The Need for Change: A New Model for Community Forest Management in Protected Areas

The problems of the Rajaji area are but a case study exemplifying the failures of existing policies for protected areas across India. These problems are inherent in the structure of a system with limited resources, which has served to alienate the local people whose support it requires. There can never be enough restrictions and enough enforcers as long as the system is widely viewed as illegitimate by local peoples. Local peoples have ceded their traditional sense of responsibility for the forest to the authorities, and respond to the officials' activities with a mixture of anger and resignation. Structural change would benefit forest officials in carrying out their duties and support the traditional lifestyles of local people. For several decades past conservationists have argued that the government managed parks protection model, which excludes local people, would be the only means to insure protection of the ecosystem. That model is in fact bringing about the destruction of the ecosystem it was charged to save. It is destroying the culture and lifestyle of the indigenous people living in and around protected areas. An alternative approach is urgently required.

Joint Protected Area Management (JPAM) represents one alternative response to the existing model. In this model the Forest Department and local peoples work as partners in evolving co-operative approaches to achieving protection of the ecosystem and meeting the needs of local peoples. It is to evolve through the existing management structure. The Rajaji area again provides a case study illustrating the problems with a JPAM approach. It is impossible to visualise, given the past fifteen years of conflict and the enormous discrepancy in powers, that forest officials and local peoples could sit around the table with a sense of equity and partnership under the structural framework of the present system. The JPAM concept of an honest partnership under the existing system is not a viable option in many settings. There must be a structural change in powers and systems in order to facilitate an effective community management model.

Community Forest Management in Protected Areas (CFM-PA)

CFM-PA is a set of organisational structures and processes for defining and managing protected areas, in which the local peoples who have traditional rights to the area become the leaders in managing the resource within a formal policy framework, while government departments and other stake-holders have a monitoring and supportive role. The overall goal is to protect and administer in a sustainable manner the ecosystem, its wildlife, and the traditional rights of the local peoples. CFM-PA requires a clear change in the roles, structures and processes of protected area management, and defines local peoples with traditional lifestyles as integral to it.

There are multiple reasons to argue for the CFM-PA approach. It can provide effectiveness in forest protection at low cost, respect for the rights and needs of indigenous peoples, protection from vested interests, and the ability to facilitate community development and environmental consciousness. Ten principles are to be implemented in relation to the characteristics and needs of the local communities and ecosystems:

- Ecosystem Protection
- Participatory, Democratic Structures
- Open Communication
- Management Responsibility and Benefit Sharing in Relation to Traditional Usage
- Gender Equity

- Community Responsibility
- Effective Conflict Resolution

- Traditional Rights and Use
- Discrete Jurisdictions and Explicit Agreements

- Effective Monitoring and Advocacy

The Rajaji Area as a Test Case for CFM-PA

The Rajaji context, and particularly the Van Gujjars who are the primary users of the core area, have the specific community characteristics which have been associated with success in community management efforts outside protected areas in India. They are a homogeneous tribal group with strong social and ecological values and a close-knit society. They have clearly defined areas of forest use both as families and a community, plus a traditional and effective decision-making structure to enforce regulations. They are specifically motivated, and request that they be given the opportunity to manage the forests.

CFM-PA within the Rajaji context also means giving lead responsibilities to local villagers, in relation to the resources and areas where they are the primary users, particularly in the border areas. They also exhibit positive characteristics for community management. They are generally subsistence peoples with a high level of dependence on specific minor forest produce resources. There is little overlap with the Van Gujjars, and there is clarity in traditional uses and respect for each other's rights where these overlap. They are motivated to take responsibility for management of minor forest produce resources and are explicitly requesting involvement, particularly in the initial pilot zones. Overall, community forest management in the Rajaji area represents an excellent test case for CFM-PA.

Objectives of Community Forest Management

Community forest management in the Rajaji area would aim to protect the ecosystem and wildlife of the Shiwaliks, and the traditional rights of the Van Gujjars and local villagers. They would have the choice to permanently live in and around the protected area in an environmentally and economically sustainable manner. These objectives would be addressed through five complimentary strategies.

Strategy 1: Core Area Community Forest Management Structure

The Van Gujjars are proposing to be the lead managers of the core areas, working in co-operation with the villagers, who would take responsibility for the bhabbar resource they utilise in the area. There would be an agreed policy framework, with monitoring and support from existing agencies such as the Forest Department, RLEK, and police officials. The Van Gujjars have lead management responsibilities in these areas because they are the only residents and forest dwellers, and have a high level of dependence on these forests for all aspects of their survival. The core area management structure provides for villager roles in relative to their levels of traditional use.

There would be a detailed management agreement enforced by a Regional Committee which would restrict use of forest resources to minor forest products traditionally used by Van Gujjars or villagers. It would require the on-going use of these products to be in keeping with the traditional lifestyle of Van Gujjars and local villagers. Commercial interests could not claim rights to use the protected area under community management, neither could Van Gujjars nor villagers change the nature of their uses for commercial purposes.

The proposals provide for a decentralised decision-making structure. The definition and implementation work is at the family and dera level. Each family would make a proposal for the number of buffaloes they would keep and the means through which they would improve their forest over five years. This proposal would be reviewed and approved by a local Khol Committee consisting of at least one adult from each dera in the khol. At least one-third of the Khol Committee

11

would be women. The day to day decision-making power is at the khol committee level, including the ability to re-apportion forest in the khol in line with the number of buffalo. Decisions would be made by the consensus of all members present. The khol committee would restrict the overall number of buffalo in the khol to a level set by a Range Panchayat. The panchayat also provides co-ordination and a conflict resolution mechanism at the range level. It is based on the traditional system of respect for the decisions of a council of elders. The range panchayat would also be responsible for regular report to the Sanctuary Committee and the Regional Committee on the problems and accomplishments under the community management structure, seeking their assistance where appropriate. The Sanctuary Committee is a structure to co-ordinate activities across ranges, and made up of Van Gujjars, villagers, the DFO forester for the sanctuary, and a training facilitator.

Monitoring and advocacy would occur at the regional level through an overall Regional Committee, which would be responsible for supervising implementation of the community management structure and policy framework, and for assuring that there is open communication and healthy problem solving between various groups. In addition, the committee would adopt an advocacy role with government to support community forest policy. The Regional Committee would also be allotted funds to hire an individual responsible for evaluation and documentation of the community management process and results.

The Regional Committee would include three Van Gujjars from the area, three representatives from Villager Minor Forest Produce Committees, the Director of each of the three sanctuaries, the Conservator for the Shiwalik Forest Division, the Chairman of the NGO (RLEK), a representative from the Wildlife Institute of India, a district Police Superintendent, and three eminent persons, including one advocate, with no vested interest in the Rajaji area. The Regional Committee would make a full public report to the Chief Conservator of Forests every year on the progress and results of the project, ensuring that all elements of the community management structure are performing within existing legislation.

There would be a comprehensive three year training component to train the committees, and facilitate communication during the implementation. It would include a training facilitator for each range, and formal training programs for all levels of the community management system. There would also be a DFO forester for each sanctuary, who would be a resource person and observer in the khol committees as well. The DFO forester would facilitate a full forest inventory carried out by the khol committee at the beginning and end of the five year management term. The community management approach would be automatically renewed if forest quality were maintained or improved over the period.

Strategy 2: Villager Minor Forest Produce Committee Structure

The traditional rights and areas of use by each village for each khol in the proposed park area to come under community management would be defined by a consensus of the relevant villages and the Van Gujjars living in the area. A Minor Forest Produce Committee would be developed, through which each village would manage its traditional forest resources and border areas. Each man and woman from the village traditionally using forest resources in the proposed park would be eligible for membership in the Minor Forest Produce Committee (MFPC). A minimum of one-third of the members would be women, as they are important users of minor forest produce. Elections would be held for an executive body of approximately twenty people for each MFPC, and the village pradhan would automatically be a member. Two Van Gujjars from the relevant Khol Committee would participate in MFPC meetings, but would not be involved in decision-making. Committees in villages which extract bhabbar grass from a specific khol in the core area would be responsible for developing a minor forest produce management plan for this resource. The plan would specify the amount of resource use, extraction procedures and regulations, planting and conservation efforts, and the supervision system. The plan would be submitted to the relevant Van Gujjar Khol Committee for its review and endorsement at a joint meeting. The relevant village pradhan and one other member of the MFPC would par-

13

ticipate in the Khol Committee with respect to minor forest produce.

Some village MFPCs have traditional areas for grazing and firewood collection in border areas of the proposed park where there are no Van Gujjars. In these areas, the MFPC would be responsible for developing a border area management plan for minor forest produce. The plans would recognise the Van Gujjars' traditional lopping rights in these areas. The village MFPCs would report to the Sanctuary Committee and Regional Committee on a regular basis.

Strategy 3: Community Forest Protection Structure

A community forest protection system would be developed for the area, in which Van Gujjars are trained to take on front-line protection duties with back-up support from the police and forest officials. Three Van Gujjars per khol would participate in a training process to become community forest guards. They would staff boundary checkpoints with Forest Department guards and would also move across the interior area of khols for forest protection purposes. Forest Department officials would provide enforcement within the khols when requested to do so by the Khol Committee or the Van Gujjar forest guards. The Van Gujjar forest guards would report all illegal activities to the appropriate authorities. They would not have permission to carry weapons. A wireless communication set to be provided for each Khol Committee so that Van Gujjars could rapidly inform authorities if illegal activities, which require backup from law enforcement authorities are occurring.

Strategy 4: Support for the Van Gujjar Nomadic Lifestyle

There would be advocacy and support for the nomadic movement of Van Gujjars, so as to maintain the viability of this lifestyle for those who choose it, and to reduce the impact of Van Gujjars on environmental resources in the Rajaji area in the summer. The Uttar Pradesh Forest Department would issue orders to Conservators and DFOs in the Hills Region requiring officials to ensure free and unobstructed access to reserve forests for Van Gujjars during the migration. The orders would require prompt prosecution of persons who threaten or harass them with respect to this use. Two Hill Monitoring Committees including Van Gujjars, DFOs in the hills, a RLEK staff person, and villagers would be established to help facilitate the migration of Van Gujjars. These committees would work to resolve conflicts between Van Gujjars and villagers which occur from time to time.

RLEK would consult Van Gujjars moving through each major migration path to determine their needs so as to reduce problems on the migration. Support programs would be developed, based on Van Gujjar requests within a philosophy of Van Gujjar self-help and responsibility.

Strategy 5: Van Gujjar Development Priorities

An essential component of community management is control of, and accountability for, financial resources. Lopping and grazing fees would be paid to the Range Panchayat rather than the Forest Department with the initiation of the core area community forest management structure. This would provide funds to support Van Gujjar development priorities in education and health in addition to providing a small wage for the Van Gujjar forest guards. The Khol Committee would appoint a Health and Education Committee to be responsible for defining health and education programs to be provided through the tax funds of the khol. This committee would have the same number of men and women in order to provide a balance of men and women's priorities in defining programs. The committee would fund and supervise a teacher to provide education for the deras in the khol.

Phases and Costs of Implementation

It is important to introduce a management change of this magnitude on a gradual basis so as to identify strengths and learn from problems as the process evolves. The proposal sets out the detailed specifications and khols and MFPC boundaries for implementation in the first year pilot zone and then recommends implementation in three additional areas on a sequential yearly basis to bring all areas under community management.

The on-going management model would be less expensive for government than the existing system as the number of forest officials required would be lower as community members would be taking on new roles and responsibilities. There would be significant short term expenditure for the training component which would be arranged and obtained through the NGO.

Conclusion

This report defining Van Gujjar proposals for Community Forest Management in Protected Areas (CFM-PA) represents a first step in a new approach to protected area management which protects both the ecosystem and the indigenous people living within it. Community management must necessarily proceed on a case by case basis as structures and approaches would vary depending on the characteristics and needs of the local communities and the environment. Community forest management must also be an evolving process as local peoples work together with resource persons and organisations. Community forest management will bring a more harmonious and positive interpersonal environment for Van Gujjars, villagers, and forest officials living and working in the Rajaji area. It will allow everybody to participate in a system in which incentives and values foster forest protection and personal respect.

PART -ONE

PROTECTED AREAS AND INDIGENOUS PEOPLE

Chapter 1
The International and National Policy Context

"It has to be done, but let others do it: that is the attitude of the wealthy, developed countries regarding conservation of global natural resources. The overall global balance has to be maintained so they are pushing the Southern countries to adopt their own model of policed conservation in "nature parks". This is because the North needs to keep its own lands and biological resources free for procuring bountiful production of food, timber and export - crop, which they can then export to the less developed countries."

- David Wood

The Rajaji case has unique characteristics which are specific to the Shiwalik ecology, the Forest Department practices and structures, and the local people living there. The policy problems are not however, an anomaly, but are similar to those being confronted throughout India and world wide. The national and international policy context is described in order to analyse the macro level factors which have influenced the local issues. The Rajaji problems have not resulted simply from the actions of individuals and local organisations. Many of the local constituencies have in fact, been forced to react in some way to these macro level policy pressures. The broader analysis provides valuable insights into past failures, and examples of successes where environmental protection is coupled with respect for the rights of indigenous people.

Protected Areas and International Conservation Policy

There are more than 7000 protected areas world-wide in over 130 nations, with the vast majority of these areas established in the last 25 years (McNeely, 1992). Dating from ancient times, specific locales have been protected, either as sacred groves preserved for religious reasons, or as recreational and aesthetic parks for royalty and the nobles. In India, Emperor Ashoka's fifth pillar gave protection to fish, animals and forests (Jena, 1993). The modern impetus for protected areas however began with the designation of Yellowstone National Park in the United States in 1872. The rationale was to protect exceptional areas for society, based on their natural beauty and uniqueness. From this point, the movement for establishing protected areas has been led by conservation interests, and subsequently government policy makers in developed countries, particularly in North America. Although the establishment of American parks was extremely controversial even in its own context, the United States did not face the problems of a large local population already inhabiting the designated areas. Most of the indigenous population had been previously exterminated and/or shifted by military intervention and colonisation. The first national park in Asia was the Corbett National Park in India, established in 1935 under British rule. It followed the pattern of providing a recreational and hunting sanctuary for elite members of society.

The environmental movement in Northern countries in the late 1960s added a new impetus to "wilderness" conservation efforts and was an important force in the rapid expansion of protected areas world wide over the next quarter century. Environmentalists demanded protection of ecological areas in the face of increasing evidence of global environmental problems, and shrinking natural resources. These environmental problems were primarily due to Northern industrialisation, consumption, and exploitation of natural resources. While steps for environmental protection were initiated in Northern countries, they were hindered by the strong vested interests of corporations and governments in the continued exploitation and consumption of natural resources. Governments and large conservation organisations began to shift energies to protecting species and areas in other parts of the world. Images of exotic, disappearing mammals such as the tiger, panda bear, elephant, and whale captured the international media's attention, and fuelled the rapid growth of international conservation organisations such as the World Wildlife Fund. Yet past colonial exploitation, and present industrialisation controlled by the international and national elite was rapidly destroying natural resources and creating environmental problems in developing countries.

As a result, the International Union for the Conservation of Nature and Natural Resources (IUCN) was formed, and conservation became an objective of existing international organisations. Western governments were simultaneously paying lip service to demands from the general public for environmental protection and conservation, and supporting the continued exploitation of natural resources in the developing world to fuel their economies. The expansion of protected areas served both ends as they ostensibly protected exotic species and provided small, high profile examples of conservation. Yet they also deflected attention from the continuing commercial exploitation of enormous tracts of land in developing countries. Protected areas in developing countries were most frequently established in areas inhabited by indigenous and tribal people. This was partly due to the fact that they did not have the political power of elite and private interests to oppose the protective designation, and partly because their lifestyles and cultural practices had preserved the local ecology.

The development of global concern with respect to the issue of biodiversity further propelled the emphasis on establishing protected areas in the 1980s and 1990s. Species extinction is a global crisis and it is estimated that more than 1000 species a year are disappearing from the planet (Wilson, 1988). 48% of the world's plant species inhabit forest areas and it is estimated that 90% of these areas will be destroyed within two decades, resulting in the loss of about a quarter of these species (Raven, 1988). India in particular is a "mega-diversity" country with 45,000 recorded wild species of plants and 81,000 species of animals, or 6.5% of the earth's known wildlife (Ministry of Environment and Forests, 1994). It is estimated that 10% of these plants are threatened, and 21% of mammals (ibid.). It is no coincidence that India's vast biodiversity is paralleled by a vast cultural diversity, with more than 4500 distinct ethnic communities and 325 languages (Singh, 1992). The cultural diversity has mainly developed out of human adaptation to diverse ecological systems, which in turn have maintained distinction in relation to the diversity in cultural practices.

International efforts and organisations focused on wildlife conservation with no mention of indigenous people. This was translated at a policy level into steps to evict indigenous people from protected areas throughout the world. Local people were seen as enemies of wildlife. For example, World Wildlife Fund-India (WWF-I) annual reports highlight species protection without mentioning indigenous people as late as 1989 (WWF-I, 1989). More recent WWF reports shows some emphasis on people's participation.

The biodiversity crisis became a focus for the "United Nations Conference on Environment and Development" in Rio de Janeiro in 1992 and resulted in the finalization of the global "Convention on Biodiversity." The convention emphasises continued priority on the establishment of protected areas and does not include the role of local communities in their establishment or management (Article 8; Krattinger et. al., 1994), despite some support for indigenous knowledge. Moreover, the conservationist focus on biodiversity has been coupled with an appeal for biodiversity in terms of the economic interests of global decision makers. McNeeley et. al. (1990) emphasises that biodiversity advocates must frame the issue relative to the economic value of biological species. There is no mention of the role of global consumption and international trade in diminishing biodiversity, particularly as it benefits the Northern countries (Gray, 1992). The result has been an international conservation policy which promotes the preservation of high profile "sanctuaries", while ignoring the global economic factors that have caused the destruction of ecosystems outside these sanctuaries. Simultaneously, given the need to protect the sanctuaries, indigenous and local people have been evicted from them and/or lost their traditional rights over these lands. The Global Environment Facility (GEF), an international green fund established through the World Bank, has taken on the role of providing funds to developing nations in support of this policy of establishing protected areas that reduce or eliminate human interference by local people.

The destruction of vast tracts of forest inhabited by indigenous people over the past three decades through commercial exploitation and encroachment, coupled with protected area policies which evict people and remove customary rights, has predictably spurred the organisation of indigenous people to defend their rights and fight for their cultural survival. In many instances protests have resulted in violence as their verbal pleas have been ignored. Tribal people have forcefully argued that their identity, cultural survival, and economic livelihood are intimately linked to the ecosystems in which they live. Documents such as the *Charter of the Indigenous Peoples of the Tropical Forests* (International Alliance of the Indigenous-Tribal People of the Tropical Forests, 1993) assert that their traditional knowledge and lifestyles have been a primary factor in the preservation of forest resources. Moreover, as the original human inhabitants, they declare their ownership and management rights over these areas. Articles 41-43 clearly state the stand of indigenous peoples regarding conservation, biodiversity, and protected areas:

Article 41: Conservation programmes must respect our rights to the use and ownership of the territories we depend on. No programmes to conserve biodiversity should be promoted on our territories without our free and informed consent as expressed through our representative organisations.

Article 42: The best guarantee of the conservation of biodiversity is that those who promote it should uphold our rights to the use, administration, management, and control of our territories. We assert that guardianship of the different ecosystems should be entrusted to indigenous people, given that we have inhabited them for thousands of years and our very survival depends on them.

Article 43: Environmental policies and legislation should recognise indigenous territories as effective "protected areas" and give priority to their establishment as indigenous territories.

The International Labour Organisation's (ILO's) "Convention Concerning Indigenous and Tribal People in Independent Countries" and the draft "Declaration of Indigenous People's Rights" at the United Nations both recognise communal rights to land and territories. Article 14 (1) of the ILO convention states "The rights of ownership and possession of the people concerned over the lands which they traditionally occupy shall be recognised" (ILO, 1989).

Conservationists respond to these assertions by stating that despite the legitimacy of indigenous rights, the exclusive approach of tribal people remains anthropocentric. Wildlife has fundamental rights as well (Kothari, 1995). Conservationists however, inevitably have their roots in the mainstream consumer society and institutions which have destroyed biodiversity and vast natural resources. Their priority has been to support protected areas funded by the exploitive, resource-dependent urban economy, rather than on supporting forest dwellers whose livelihoods are connected with the forests. In addition, given the past record of problems in protected areas with respect to encroachment and the protection of wildlife, there is much evidence to suggest that governments and conservationists cannot do as well as indigenous people in protecting ecosystems (Wood, 1995; Kemf, 1993). In the 1990s there has been increasing attention and progress in establishing the rights of indigenous people and demonstrating their success in managing protected areas (Kemf, 1993).

International NGOs in recent years have strongly taken up the cause of encouraging the involvement of local peoples in forest and protected area policy. Social justice and environmental movements in both the South and North have put pressure on large multi-national donor organisations such as the World Bank to emphasise local involvement and joint responsibility in forestry programs. Consultation and local participation are now buzz words in World Bank and Global Environment Facility policy statements, but even internal evaluations point out major deficiencies in practice (Johnson, 1994) while external critics identify the continued extraction-based free market economic model and top-down planning process as continuing to dominate funding programs. For example, in Thailand, international agencies look to fund the commercial and government interests which are now espousing a green philosophy while continuing to ignore the village-based environmental movements which have been effective in preserving ecosystems. However, the grass roots movements are committed to local land rights and resistant to free market exploitation of resources, contrary to the priorities of international agencies (Lohman, 1992). Past policies continue in practice despite the new philosophy.

The proposal for the Rajaji Sanctuary to become a national park has followed the international trend. Hence, a historical review of Rajaji forest management practices indicates that the main emphasis with regard to wildlife until the 1960s "was on shooting" (Kumar, 1995; p. 107), and it was not until the 1970s that there was specific mention of wildlife conservation issues. Moreover, the preservation of an endangered species, the Asian elephant, remains a major reason for the proposal to make the area a national park. The specific area designated was chosen in part because it was the best protected area of a degraded Shiwalik ecosystem, and partly because it was adjacent to Dehra Dun which has a strong British and conservationist legacy. Not surprisingly, as in other locales, the area coincided with the winter home of a forest-dwelling, nomadic, tribal people, the Van Gujjars, who had been using these forests for several centuries. Finally, given the international and national policies of removing local people from national parks, a large amount of energy and resources was directed to removing the Van Gujjars from the proposed park and eliminating the customary rights of other local people. Predictably, as in other areas, these actions resulted in massive protests from both Van Gujjars and villagers; eventually the Forest Department was forced to back down from the eviction. Most recently the Wildlife Institute of India, whose staff have fully supported and been proponents of the protected area eviction policy and philosophy, has now received funding from a major international donor, the Ford Foundation, to explore ways to improve relationships between the Forest Department and local people.

To summarise, the Rajaji context parallels international trends, and has been strongly influenced by them. Remote but powerful international forces have shaped the Rajaji conflicts. Efforts to alter the local problems must acknowledge that these forces are at work. Next, it is important to examine national policies and pressures which have influenced the Rajaji.

Forest and Protected Area Policy in India

The "Statement of Shared Concern" in the second citizens report on the "State of India's Environment" (Centre for Science and Environment, 1985) places the international trends in a national context:

> Nature can never be managed well unless people closest to it are involved in its management and a healthy relationship is established between nature, society, and culture. Common natural resources were earlier regulated through diverse, decentralised community control systems. But the state's policy of converting common property resources into government property resources has put them under the control of centralised bureaucracies, who in turn have put them in the service of the more powerful. Today, with no participation of the common people in the management of local resources, even the poor have become so marginalized and alienated from their environment that they are ready to discount their future and sell away the remaining natural resources for a pittance.

It is important to examine the status of forests and protected areas in India and analyse how the present situation developed as the result of government policy and law through history. This analysis identifies additional massive factors which have brought about problems in the Rajaji area.

Approximately 75 million hectares of land are under forest cover in India (23% of the total area; Saxena, 1995). The quality of these forests is however, poor. More than half these lands have less than 40% crown cover (World Bank, 1993). Although there has been a large increase in recent decades of the total amount of land under control of the Forest Department (a 50% increase between 1960 and 1980; Saxena, 1995), there has been an on-going degradation of these government managed forests. S.A. Shah (1995), a retired Chief Conservator of Forests from Gujarat summarises the ecological status of forests in a detailed report on the "Status of Indian Forestry" as:

> India's forests are ecologically unstable and unhealthy. The process of conversion largely through clear cutting has destroyed the primary structure of most of the forests. Repeated

fires and overgrazing have altered the ground and soil structures substantially. (p. 21)

The fundamental policy issue is the degradation and dis-
appearance of existing forests rather than the official loss or
transfer of existing forest lands to other uses. This trend has
occured in parallel with increasing government control and
jurisdiction.

There is an identical pattern of increasing abundance and
increasing degradation in lands accorded protected area
status. Ministry of Environment and Forest statistics indicate
that India has 441 sanctuaries and 81 national parks with a
target of more than 750 protected areas (Bhattacharyya, 1995;
SPRIA & RLEK, 1993). The majority of existing protected areas
are not officially established because local people hold outstanding claims and rights on their lands, which must be settled and vested in the state prior to official proclamation
(Indian Institute of Public Administration, 1994). Table 1 demonstrates the dramatic increase in sanctuaries and parks since
1960, paralleling international policy trends (SPRIA & RLEK,
1993). 4.3% of the total land of the country is accorded protected area status, and approximately two-thirds of these
lands are covered by trees, representing 20% of all forest lands
(World Bank, 1993).

A report on the status of parks and sanctuaries (Kothari et.
al., 1989) indicates that 69% of the protected areas responding
to their survey state officially that they have human populations residing within their boundaries. Kothari et. al. (1989)
estimate that approximately three million people live in protected areas, with a far higher number in peripheral areas.
Over half the protected areas surveyed indicated that a part of
their area was used and occupied by other government departments and agencies. "Legal" rights and leases were being
utilised in 64% of the areas with "illegal" usage in a far higher
proportion. These lands are home to approximately 20% of the
tribal population of the country (SPRIA & RLEK, 1993).

Table 1
Increase in Number of National Parks and Sanctuaries In India

Year	No. of Sanctuaries	No. of National Parks
1960	60	5
1975	12	5
1985	247	53
1989	411	69
1991	421	75 (18 Tiger Reserves)
Future	633	147

Excerpted from SPRIA & RLEK (1993), page 2.

These patterns are not surprising given India's high rural and tribal population, and the fact that many protected areas were only established within the last two decades. A Ministry of Environment and Forests report prepared for the United Nations Development Programme notes that "these [protected] areas are poorly managed with little consideration given to the local people living in and around them" (Indian Institute of Public Administration, 1994; p. 5). Reports indicate that rampant poaching and industrial encroachments are threatening protected areas, including the high profile tiger population in Madhya Pradesh, even though Madhya Pradesh was recently declared a "Tiger State" (Ral, 1995). Inadequate government resources and forest officials' corruption are noted as primary issues (John, 1995).

Grazing has been identified as a particularly important biotic pressure on forests and protected areas which has increased over the years (Kothari et. al., 1989). Government policy has focussed on ways to reduce or eliminate this grazing pressure by restricting the rights of local peoples and attempting to wean them off free range cattle (Indian Institute of Public Administration, 1994). However, Saxena (1995) estimates an increase in the livestock population of 1.2% per year between 1951 and 1987. During the same period human population increased 2.1% per year and the production of food grains 2.8% per year. He argues that if forest productivity had increased, overall grazing pressure would have been stable or even reduced. In fact, grazing pressure has increased primar-

28

ily due to deterioration of the forests rather than the increasing demands of local people.

Despite the increasing degradation of forests and protected areas, the loss of rights and oppression of local people, and the increasing parks and people conflicts, protected areas have fulfilled the function of severely reducing the ability of urban and business interests to swallow up these lands and resources for financial gain. Without protected area legislation, there would be little to stop powerful interests from buying out or simply pushing out local people. Instead, commercial and industrial encroachment require denotification by state legislatures, which at least expose them to public scrutiny. Although denotification has been occurring more frequently in the past few years given the nexus between business, criminal and political interests, the resulting public debate has protected lands in some instances (Bhattacharyya, 1995, Kothari et. al., 1995). In addition, government and commercial encroachment into protected areas occur under the table in some instances, despite legislation. Local people find themselves caught between commercial concerns which will evict them with denotification and development, and Forest Department and conservationists who will evict them or undermine their lifestyles through wildlife protection measures. The net result is the degradation of protected areas and the exploitation of local people. Parts two and three of this report document this process with respect to the proposed Rajaji National Park.

The present deterioration in forests and protected areas in India and exploitation of local people is part of an international trend as previously described, yet it is also unique, due to India's colonial and historical legacy which bears specific attention.

History of Forest and Protected Area Policy in India

The present status of forests and protected areas can be attributed to four factors: the government policy of exploitation and use of forests for commercial revenue generation, the resulting alienation of local people from taking responsibility for the forests, coupled with their increasing need for forest produce for survival, the increasing demand of urban centres for forest products, and exploitation and destruction by industrial interests (Saxena, 1995). These factors have been evident throughout the modern history of forest and protected area policy in India.

There is a long history of forest protection and sustainable use in India prior to British times. Indigenous people's life-styles were compatible with a sustainable environment rich in both biological and cultural diversity. Sacred groves were common as a part of their spiritual traditions. Among the elite, both Hindu and Muslim rajas established protected areas for hunting and recreation (Kothari, et. al., 1995). The land was protected through social mores and customary rights rather than by a code of law. These relationships and practices were fundamentally and permanently altered by the advent of the British.

The primary reason for colonisation was the British inter-est in using Indian natural resources to fuel its economy. At the same time, they introduced their own legal concepts and framework, including reliance on individual land ownership and the rule of law with courts to determine rights and resolve disputes. These laws and rights were decided upon for and by British interests. The major use for forests was to provide wood to sustain railroads, industry, and ship building. Major tracts of forest were clear-felled with no thought for conserva-tion. Private land ownership was defined and taxed, and large tracts of forest which had never been owned in the British le-gal sense were claimed as government property. If the British had recognised traditional land rights, the government would have lost access to these resources while local people would not have been obliged to pay taxes on them (Singh, 1989). Communities and individuals who did not register, or were not aware of the need to claim formal land title, found that their lands were now defined as government owned. A memorandum in 1855 first restricted forest dweller's rights, and the first Forest Act passed in 1865 gave the government the right to declare any land covered by trees as Government forests and manage them as it saw fit. The Indian Forest Act of 1878 proceeded to differentiate these forests into reserve for-ests, protected forests, and village forests and to define timber duties, which became a major source of government revenue (Hiremath et. al., 1995). In 1927, the government passed a comprehensive forest act, still in force today, which contained all the major provisions of the previous acts. It continued to emphasise the other revenue yielding extraction uses of forests.

Elite members of British society also had a strong interest in wildlife, conservation and hunting through preserved areas. These recreational areas already existed under the rajas and the British continued to support this concept. India's first wildlife protection law respecting elephants was passed in 1873. The initial National Park Act was passed in 1934 followed by the proclamation of Corbett National Park the following year. Despite independence and a change from the ruling British to an Indian elite, the policies of forest extraction and protected area preservation for elite interests continued with few changes.

The National Forest Policy of 1952 placed more emphasis on ecological and social issues in forestry but ultimately noted that the forests could not be used at the cost of national interests, typically defined with respect to the defence and industrial sectors (Saxena, 1995). The subsequent implementation remained focused on the ability of the forests to produce revenue through extraction. India's first Prime Minister, Jawahar Lal Nehru, symbolised the twin policy interests. Nehru, a son of one of Northern India's elite families, was committed to using the forests through a centralised planning process for national needs including the development of an independent socialistic economy. Yet he also was an avid naturalist who set up the Indian Board for Wildlife (Kothari et. al., 1995). Indigenous people were not a priority of this centralised system yet, since protected areas were limited in number and area, urban demands on the forest were lower and forest resources more abundant, there was less frequency of conflict with rural people.

The modern environmental movement gathered strength in North America and Europe in the 1960s. This brought international pressure to bear on India to increase protected areas in line with the North American model of fencing off pristine wilderness areas and protecting them from human intervention other than that felt to be compatible with conservation interests. India attracted particular attention because of its diverse and high profile exotic species (tiger, elephant, and rhinoceros). The passage of the Wildlife Act of 1972 and the 1976 report of the National Commission on Agriculture (NCA) were the culmination of anti-people approaches to forests and protected areas. The former protected a small percentage of high profile forest lands for conservation while the latter reinforced the philosophy of exploitation of the vast majority of the forests for revenue and commercial interests.

The NCA report stated that "Production of industrial wood would have to be the *raison d'être* for the existence of forests. It should be project-oriented and commercially feasible from the point of view of cost and return" (Government of India, 1976, p. 32). The report also recognised the increasing deterioration of village lands and the increasing local pressure on forests. It blamed the rights and practices of indigenous people for the destruction of the forests. The proposed solution was to grow trees on village lands in order to take pressure off the forests. But problems in project design, decision-making processes, and difficulties in implementation, resulted in the failure of social forestry to achieve the objectives of providing more fuelwood and fodder for rural people (Saxena, 1995). Simultaneously, in 1976, forests were transferred from the state to the concurrent list of responsibilities in the Indian Constitution, becoming a joint state/centre responsibility. The National Commission on Agriculture was also the impetus for the draft Indian Forest Bill of 1980 which followed up on the need to protect the commercial value of the forests from the destructive practices of local people. Opposition to this bill brought together for the first time a wide range of local people and activists who had taken up the struggle against increased exploitation and loss of traditional rights in the 1970s. The draft bill was defeated, and demands for a more people-orientated approach gathered strength.

The 1972 Wildlife Act, amended in 1991, was the other branch leg of the anti-people approach to forest areas (Natraj

Publishers, 1992). While the forests were to serve commercial interests, protected areas were legislated to emphasise the government's environmental and conservation commitment. The Act fuelled a dramatic expansion of the protected areas' network. It provide uniform protection for wildlife, prevented hunting and trading, and defined the process for establishing and managing sanctuaries and parks. There is no requirement of public participation in defining and managing protected areas. A person must make a written declaration to claim a prior right in the protected area within two months of the government's official announcement creating that area. Local people in fact were rarely informed of the proposed protected area, did not have any conception of its potential impact on their lives and rights, nor the understanding to file a written claim for their rights. Moreover, given that indigenous people had lost their traditional rights through government usurpa- tion over the past century, many people had no legal rights to declare. In addition, the Act did not recognise the traditional practice of collecting minor forest produce as a "right."

The Wildlife Act specifically restricts access to protected areas, excluding local people except at the discretion of the warden. The warden is also able to place restrictions on graz- ing and any other activity which might damage wildlife or wildlife habitat in a sanctuary, while all grazing is prohibited in a national park. National parks are defined such that no one can retain a right over land in the park. The process for claiming and settling rights leaves an immense amount of dis- cretion in the hands of the Collector who is charged with re- sponsibility for settling rights and providing compensation.

In brief, local communities, which had been entirely de- pendent on the surrounding forest area for generations, faced severe restrictions or eviction from their surroundings. This was the final step in a long process of local people losing psy- chological as well as practical ownership and sense of respon- sibility for the land. The refusal to recognise their traditional rights, years of destructive government practices, and increas- ing population pressures have resulted in alienation of local people from the forests that had been their traditional respon- sibility. Moreover, the sense of alienation and insecurity breeds a pattern of using the forests to the greatest extent possible, given the fear that their access may be completely eliminated in the near future.

The "National Workshop on Declining Access to and Control over Natural Resources in National Parks and Sanctuaries" brought together activists and local people from all over the country with unique case studies fitting within this framework of exploitation (SPRIA & RLEK, 1993). The workshop report highlights more than twenty case studies from among many others across India in which local people have been deprived of their traditional rights and livelihood. Betla Tiger Reserve in Bihar includes seven villages in its core and 105 in its buffer zone encompassing 65,000 people and 45,000 cattle. The Phoolvari Ki Nal Sanctuary in Rajasthan has 30,000 inhabitants, while Sitamata Sanctuary in the same state includes 42 villages. The pattern is similar in all corners of the country (i.e., Similipal Tiger Reserve in Orissa, Hingolgarh Sanctuary in Gujarat, Mudumalai in Tamil Nadu). Local peoples were prohibited access to the forests or obliged to pay officials to maintain illegal access. It is estimated that forest officials extract Rs. 8,000 per year from one village of 52 families in the Sariska Tiger Reserve (ibid.). Forest officials often rule the lives of local people with arbitrary authority. Rural development programs and facilities are not accessible, as these villages are not sanctioned or officially designated on these lands.

The Wildlife Act prevents local people from taking any action against wild animals who damage and destroy crops and in some instances kill people. Compensation is small and procedures are so complicated that local people never actually receive what compensation is due to them (Pottie, 1995). Compensation for the death of a child is Rs. 5000 and for the death of an adult Rs. 10,000 in Uttar Pradesh, whereas punishment for offences against wildlife may include fines of up to Rs. 25,000. In fact, deaths due to wildlife are not uncommon and crop destruction is rampant. According to government census figures, 19 people were killed in the Almora District of Uttar Pradesh in a three year period while a non-government census indicated 35 deaths (RLEK, unpublished). In contrast, fines for illegal activities in forests outside protected areas are minuscule, set by the antiquated 1927 Indian Forest Act.

Anger and isolation has bred agitation across the country among local people. More than 20% of the protected areas surveyed in 1989 reported having had physical clashes with local people (Kothari et. al., 1989), and one would expect significant under-reporting from higher level officials on such an issue. Seven villagers were killed in protests in one sanctuary (Kothari et. al., 1995). Moreover, the implementation process, of determining exactly which areas should be protected and what status they should be accorded, has been arbitrary and flawed. In some places, including the famous Himalayan "Valley of Flowers", bans on traditional grazing have had a negative impact on diversity and required a re-introduction of grazing or grass cutting (Bhatt, 1993b; Vijayan, 1991).

The activism which resulted from the anger of local peoples and the mounting evidence of forest and protected area deterioration brought a change in official thinking in the mid-1980s as government and forestry officials acknowledged that forests could not be protected without the support of local people. The Forest Policy of 1988 (Ministry of Environment and Forests, 1988) represented the first real shift in policy in a century.

It states that:
> The principal aim of Forest Policy must be to ensure environmental sustainability, and maintenance of ecological balance (including atmospheric equilibrium) which are vital for sustenance of all life forms, human, animal and plant. The derivation of direct economic benefit must be subordinated to this aim.

Its objectives for the first time included:
- Meeting the requirements of fuelwood, fodder, minor forest produce, and small timber of rural and tribal populations.

- Creating a massive people's movement with the involvement of women, for achieving these objectives and to minimise pressure on the forests.

It emphasises the importance of the traditional rights of forest dwellers and the importance of their role in protecting the forest. Priority for the use of forest products is given to forest dwellers and the use of the forests for industrial purposes is discouraged. It de-emphasises plantations in favour of mixed and natural forests.

Following up on this new forest policy, the Government of India issued a set of Joint Forest Management Guidelines in 1990 encouraging Forest Departments to involve local people in the management of the forests. Over the past five years, fifteen states have in turn issued their own guidelines for joint forest management. There has been much experimentation and some success in this area with respect to village and reserve forests (see chapter two). There have also been obstacles as there are bound to be to any major shift in policy (Saxena, 1995). The new Centre policies are not backed by law and have to be carried out by state governments which do not necessarily share their perspectives. There are very powerful commercial concerns both within and outside all governments who want forest and protected area land for industrial development. Secondly, there are very limited funds for what is potentially a major reform of past practices and structures. Finally, and most important, Forest Departments and officials were trained and have worked under the same mandate and set of interests for a century, resulting in much resistance to a shift of this sort.

Although the 1988 Forest Policy brought a fundamental policy shift for forests in general, there was no parallel shift for protected areas as they remained under the framework of the Wildlife Act. In fact, the 1991 amendment to that Act increased restrictions on the use of a sanctuary for local people and made the final notification process simpler. The alienation of people from protected areas increased despite the change in overall forest policy.

The past decade has finally brought an increasing consensus, except among vested commercial and bureaucratic interests, that past policies have been contrary to the interests of the forests and the people living in them. After 150 years of struggle, there is now some hope that national policy could help make positive changes in local contexts, at least outside protected areas. The recent focus has been on examining new and more productive alternatives which involve local people. Uncertainty remains however, as the recent draft Forest Bill to replace the 1927 Forest Act backtracks on the positive initiatives of the 1988 Forest Policy and proposes to return India to traditional commercial approaches. Within protected areas there is a policy vacuum as the same consensus is developing with respect to the failure of present approaches, and yet the Wildlife Act continues to enforce an anti-people approach. Forest Departments have not taken any initiative to involve local people.

To summarise, national policy has generally followed international policy trends, resulting in even more pressure on local constituencies to take action increasing ecological degradation and parks–people conflicts. Yet there have been some progressive developments of late due to pleading by local people and activists at the national level. The next chapter reviews experiences in the development of policy alternatives involving local people over the past two decades. This work can provide valuable lessons in the development of a model for Community Forest Management in Protected Areas.

Chapter 2
Alternatives for Local Involvement
and Forest Protection

"Government should promote and provide opportunities for the participation of interested parties, including local communities and indigenous people, industries, labour, non governmental organisations and individual forest dwellers and women in the development, implementation planning of national forest policies."

Rio Earth Summit Communique, Principles/Elements 2(d), 1992

The review of alternative approaches for local involvement in forest protection is sub-divided between experiences in village and reserve forests, and work in protected areas. The key difference between these forest types is the difference in the legislative framework (the more flexible 1927 Forest Act versus the more restrictive 1972 Wildlife Act) and the resulting differences in the amount of practical Forest Department authority. There has been an inverse relationship between the frequency of local management experiments and the amount of control that the Forest Department has over the forest area. Thus, the highest frequency of innovation has come for "village" forests, often wholly initiated by local communities with or without Forest Department approval and support. These forests are often degraded and have little value to the Forest Department. The Forest Department has not had the practical resources to manage them in recent years. There has been a moderate level of experimentation with local alternatives in reserve forests, where the Forest Department has a significant presence and investment in resources. Finally, there has been practically no experimentation in protected areas managed under the Wildlife Act, where the forest bureaucracy is strongly entrenched. There are similar pressures, forest conditions, and human populations in all three types of forest areas, despite differences of degree. Thus, the experiences in village, protected and reserve forests are a rich source of learning which must be examined before proceeding to review alternative approaches in protected areas.

38

Whether one is considering village and reserve forests, or protected areas, it is important to distinguish between conceptual differences in types of approaches. Unfortunately the label of "joint forest management" has been used by different people to mean very different approaches in different places, resulting in a significant amount of confusion. This chapter initially clarifies the concept of "joint forest management" into three types of approach: participatory management, joint forest management, and community forest management. Since the vast majority of experiments with these approaches in India has been outside protected areas, this experience is reviewed in order to gain insights that might be applied within protected areas. Finally, we specifically review the three proposed approaches for local involvement within protected areas. Given the very limited experimentation in India, we draw heavily on international experiences to evaluate approaches.

Clarification of Concepts

There has been widespread use of the term "joint forest management" in all sorts of contexts (Johnson, 1995). The result has been that one advocate's "joint forest management" experience or proposal bears little similarity to another advocate's proposal. It is essential to make distinctions or discussions will be confused and unproductive, particularly in relation to the degree of community versus Forest Department involvement. For the purposes of this report, the conceptual definitions proposed by Johnson (1995) will be adopted:

Participatory Forest Management (PFM) is where the Government takes initiatives, manages the resource and the community participates in various forms, most commonly as hired labour.

Joint Forest Management (JFM) is where the owner (the Government) as well as the user (the Communities) manage the resource and share the costs as well as the benefits thereof.

Community Forest Management (CFM) is the case where the community takes the lead, manages the resource while the government is a passive supporter or observer.

These three categories create an implicit dimension, and inevitably there are implementations that fall somewhere in between one or the other. Participatory management is akin to public consultation with the government officials making decisions which hopefully include local interests. Joint forest management implies a partnership of forest officials and local people within the existing Forest Department management structure. Community management involves recognising a different authority structure in which the community manages the forest. In reviewing these approaches, it is important to compare approaches and to examine the results with respect to the characteristics of particular situation.

History of Alternative Approaches Outside Protected Areas

The documentation of experiments outside protected areas has principally related to the community and joint forest management approaches, as participatory approaches are often not seen as significant departures from the status quo.

Community Management Examples

The recent history of successful community involvement in forest protection in numerous states predates the reversal in official policies in the 1980s (Agarwal & Narain, 1992). Khaksitola, a small village of Oraon tribals began protecting their forests in the 1950s through community management, by blocking timber extraction by contractors, using panchayats to decide upon forest usage, and assigning villagers to guard duties (Mukhia, 1994a). In 1978 contractors, who had been given contracts for commercial extraction on these government lands, were chased away by Oraons with bows and arrows. Today, the villagers contribute rice to their guard's salary, and the village committee decides on each tree that is cut. The village forests have regenerated to become green and lush.

Five areas in Orissa unilaterally initiated successful village forest committees in the 1970s and early 1980s following a community management approach (Saxena, 1995). Large tracts of village lands have been successfully re-forested, though villagers have had to struggle with inter-village conflict over exactly who should receive the benefits, from the forest products they have protected over the years, now that extraction is appropriate. The Chipko movement in Garhwal grew directly out of villager's protection for their forests. Although the movement prevented forest destruction by commercial interests, the resulting government attention brought protected area status to the region so that the forests are still not available for local use, the original aim of the protest movement (Mitra, 1993). Effective community-initiated protection in Bihar has arisen in response to crises due to exploitation, drought, or inter-village conflict (Mukhia 1994a). Sixty forest protection committees in a tribal area are functioning in the Sinhbhum district. Some of these efforts date back to the 1970s, and have successfully regenerated forest areas, based on rules and regulations defined through their own structures, with no involvement from the Forest Department (Singh et. al., 1995). Villages in Gujarat have also initiated strong forest protection initiatives, with disputes being settled within the community. The Forest Department has endorsed these efforts, but would now like to see them formalised under JFM guidelines. Villagers are wary of Forest Department involvement undermining their efforts (Mukhia, 1994b).

Joint Forest Management and Participatory Forest Management Examples

High profile experiments in West Bengal and Haryana in the 1970s also pre-date the official "joint-forest management" policy and in these instances forest officials were important catalysts and facilitators in the efforts and successes (Chandra & Poffenberger, 1989). The Divisional Forest Officer (DFO) in the Arabari district of West Bengal initiated discussions with local villagers, which over the years resulted in a very successful joint/participatory management scheme. Forest protection committees were established including villagers and forest officials, and villagers shared in the profits from forest regeneration in turn agreeing to restrictions on forest use assisted with enforcement. This success spurred a second generation expansion of this process in West Bengal to the point that there is now a program with 1,250 villages covering 152,000 hectares. Villagers now receive minor forest products from the forest areas and some employment from the Forest Department to maintain them. The committees appear to follow a similar set of management decisions and processes defined by forest officials. Given the magnitude of the expansion, results have varied including significant successes.

Case studies indicate that rotating volunteers as guards is more effective than paid guards as they maintain, or possibly demonstrate villager commitment. The quality and commitment of the lower level forest officials was found to be the key to success as the original initiative came through these officials (Chandra & Poffenberger, 1989). Studies show that the smaller the number of villages in a forest protection committee, the greater the number of households participating, and the greater the percentage of tribal composition, the more the likelihood of success (Saxena, 1995).

Recent reviews of the West Bengal program emphasise dilemmas in the process of instituting programs within the Forest Department structure (Sarkar, 1995). There are debates as to whether committees or foresters should determine length of cutting rotations, whether benefit sharing will lead to devolution of power from forest managers, and whether forest department working plans should overrule committee harvesting plans. A village co-operative from Gujarat to West Bengal sites found enthusiasm and commitment for forest protection in some villages, and anger and lack of empowerment in others (Wasteland News, 1995). The widespread official implementation of joint forest management in West Bengal may bring more rigid standardisation of the process and the rules. Johnson (1995) argues that the West Bengal programme has evolved into "participatory forest management." The perspectives of the Chief Conservator of West Bengal bears this out. He writes, "The primary responsibility for implementation of joint forest management lies with the Forest Department... Since it is the forest staff who will have to take up the managerial duties, they have to be adequately groomed for the purpose." (Palit, 1992, p. 58). The Chief Conservator proceeds to identify the lack of resources and institutional constraints as problems in implementing this sort of process on a broad basis.

In Madhya Pradesh, the Harda Forest Division has implemented an effective joint forest management programme involving 150 village committees covering 75% of the division with the leadership coming from the Divisional Forest Officer (Bahuguna, 1993). In addition to villager enforcement and protection, the work in this area emphasised broader ecodevelopment and rural development. A spin off of this approach has been the efforts of the district forest officer to specify a wide range of planning and silviculture techniques which will allow more community control in joint forest management processes (Rathore & Campbell, 1995).

Haryana has been a leading example of the successes of alternative approaches which have elements of joint and community forest management. Efforts were initiated in the 1970s by forest officials in the Sukhomajri and Nada watersheds (Sengupta, 1993; Gupta, 1995). This Shiwalik Hill area is ecologically similar to the Rajaji area. Dialogue was initiated with villagers by forest officials, and evolved into successful pilot projects. Autonomous registered village organisations were formed and called Hill Resource Management Societies (HRMS). Earthen dams were built for water conservation, trees planted, forest areas protected, and villagers agreed to stop grazing in the area. The HRMSs received leases for the extraction of forest products, bhabbar grass being of particular significance. Under HRMS management, bhabbar grass productivity in the Nada watershed increased through protection and planting from 36 kg/ha (wet weight) in 1981 to more than 5,000 kg/ha in 1986, with similar results in Sukhomajri. Fodder grass rose from 35 kg/ha to 600 kg/ha (Poffenberger & Sarin, 1995).

The pilot approach was extended to 38 other areas with some success, though results were not as positive as those in the pilot areas (Saxena, 1995). There were difficulties in communication between the Forest Department and the HMRSs and debates arose over resource sharing and procedures. In 1988, the Forest Department with the involvement of an NGO, began to address difficulties in policies and structures and increases in tree cover have continued in recent years.

The Haryana experiences have emphasised a higher community control and management component than in other JFM areas in that the villager management committees are fully registered autonomous organisations that cannot be dissolved by the Forest Department (Sarin, 1993). They have the freedom to define rules for their own members, and the right to seek compensation if the Forest Department does not honour its commitments. Moreover, the Forest Department has leased the HRMSs full rights to management and extraction of resources, whilst requiring specific protection measures. Resource profits are shared with the HRMS based on gross income (Sarin, 1993). In other jurisdictions the sharing is often based on net income, which can result in communities getting little or nothing once the Forest Department has taken out overhead costs.

Haryana Forest Department procedures have caused problems for the HRMSs, yet the autonomous nature of the HRMSs gives them more ability to press for changes in policies. HRMSs are working to form an autonomous federation to challenge the Forest Department on issues of concern shared across HRMSs. One problem has been that the Forest Department demands full payment for bhabbar leases at the outset, yet private contractors are allowed to pay in instalments. HRMSs do not have the capital to pay leases at the start. There have also been difficulties with inter-village conflict when working together under one HRMS. The HRMSs are now recommending that management areas be subdivided so that fewer villages manage smaller areas (Poffenberger & Sarin, 1995).

Comparisons of Alternative Management Approaches

With a range of community, joint, and participatory management structures now in place, it is important to compare different approaches. Raju (1995) studied twenty forest protection committees (FPCs) in Orissa, Gujarat, and West Bengal. The results defined an anagram for six characteristics of successful forest protection committees labelled TRAITS (see following box).

The study recommends that the constitution of the FPCs is best left to the people themselves. They need the power to enforce forest protection against smugglers, as Forest Department response time was found to be too slow to be effective. Those who have been given this power have dealt with offenders fairly and communicated openly with the Forest Department. Based on the results, Raju (1995) recommends that "The Forest Department's role should be one of facilitation and support, but it should move away from the control function" (p. 99).

Johnson argues that any one form of forest management will not be appropriate to all situations. It is potentially disempowering and destructive for Forest Departments to take a controlling role in areas where community success is likely given local characteristics. Yet one is unlikely to expect success in community management from a community which shows little initiative, homogeneity, and much conflict. A range of characteristics seems to be correlated with community management strength and success (Agarwal & Narain, 1992; Johnson, 1995; Raju, 1995; Sarin, 1995; Saxena, 1995). These include communities and structures which have:

- a culturally homogeneous make-up and set of values
- high percentages of tribal people
- lower levels of economic disparity
- high levels of dependence on the forest for survival
- consensus structures with high levels of actual participation, rather than political organisations
- clearly defined forest use areas with minimum overlap with other groups and interests
- forest protection committees high on TRAITS dimensions
- autonomous status with accountability and conflict resolution mechanisms
- greater gender equity and involvement of women
- land with capacity for regeneration rather than fully degraded land where successes will be slow and small
- systems based on equity to distribute the benefits of protection

According to Mr. G. Raju, TRAITS of successful Forest Protection Committees are :

- *Transparency:* Openness in internal and external activities, particularly involving the flow of funds, is associated with success.

- *Rules:* Rules for forest protection and use are essential, and acceptance is highest if they are self-imposed.

- *Awareness:* High awareness of the forest resources and the rules governing use and protection is important.

- *Initiative and Independence:* Self-initiated FPCs are performing more effectively and are not dependent on external agencies for support.

- *Tenure:* Successful FPCs feel the forest belongs to them based on their regeneration efforts, although they recognise that legal title remains with the government.

- *Satisfaction of Needs:* FPC well-being is based on it being established to fulfil the members' basic needs.

The research demonstrates that community management is clearly the preferred system where many of these characteristics are in place or capable of development. Simply from a perspective of conserving government resources, it makes little sense to involve the Forest Department to any degree which is not essential to success. Moreover, there is the danger that what begins as an effective informal Forest Department intervention may evolve into a rigid broad-based system which reduces effectiveness and community capacity. On the other hand, if there is a high level of support and initiative for joint approaches within the local Forest Department and an unmotivated and divided community, participatory or joint management schemes may be the most appropriate. Forest Department initiatives seem to have been effective when particular district forest officers have exerted a high level of initiative, flexibility, and independence at the outset, and subsequent institutional structures are set up to facilitate community autonomy and decision-making. A former Chief Conservator of Forests in Gujarat argues:

> The Forest Department was created and structured over 125 years ago to meet specific objectives, namely to ensure sustained yield of timber for industry and to increase revenue for a foreign ruler. These objectives are no longer pertinent. However the structure and composition of the Forest Department continue to remain largely the same as when it was created. There has been some growth but little development. Is it possible to implement a new concept with the same old institution and infrastructure? It is very doubtful. A few isolated successes are there, but they are probably exceptions... It appears to me that we are putting the cart before the horse. If we are really serious about JFM, we should have started structural changes in forestry institutions. It is not too late even now, however. (Shah, 1993; p. 38)

Individual successes initiated by District Forest Officers may in fact facilitate the structural change process within forest departments but as yet there has been little overall change. States have not followed up on the Forest Policy of 1988 with major revisions in state policies or re-structuring of Forest Departments, though changes in Haryana and West Bengal seem to be evolving slowly. It is noteworthy that most state guidelines for joint forest management target "degraded forests" indicating that community involvement is only viewed as a priority option where all else has failed, rather than considering that local involvement might also be essential to maintaining and improving other forests.

Key lessons from the experiences outside protected areas are that, if at all possible, emphasis should be placed on the community management end of the continuum, whilst recognising that innovative forest officials may be valuable instigators of change in areas lacking in community initiative and strength. Those officers, local people, and NGOs must also create pressure to facilitate the change process within the Forest Department. Inevitably institutional change is a difficult and long term process. More rapid results are likely to be achieved through capable communities adopting community management approaches. In fact, it may be more productive to examine how to facilitate community management in areas that presently lack key capability characteristics. Positive community development among weaker communities will create positive results in all aspects of community life. Community management, institutional change, and community development become priorities. There is much to be done in all three areas, but the learning which has already resulted from experiences outside protected areas can provide a foundation for examining proposals and work within protected areas.

Alternatives for Local Involvement and Forest Protection in Protected Areas

There is a similar set of alternatives for local involvement with respect to developing management options for protected areas, although the institutional context differs in that there are more legislative constraints and a stronger Forest Department presence. While Joint Protected Area Management (JPAM) and Community Forest Management in Protected Areas (CFM-PA) are extrapolations from previous efforts, participatory management approaches in protected areas have had a higher profile than outside these areas, particularly work with respect to Biosphere Reserves and Ecodevelopment. Efforts in these areas are reviewed before considering JPAM and CFM-PA.

Participatory Management

The concept of Biosphere Reserves was initiated through the United Nations in the late 1940s, but there has been far more discussion of it among the scientific community at the international level than there has been practical application at the national or grassroots level (Nair, 1993). Given the increasing conflicts between people and protected areas in the 1970s and 1980s, it gained profile through UNESCO because it appeared to be a protected area status which included humans as an integral part of the ecosystem. In India, the work of a national committee resulted in the approval of seven reserves and proposals for an additional seven (Kothari et. al., 1995). They have however, been restricted to re-naming existing protected areas which are managed under the same legislative and bureaucratic structure. Suggestions to formulate new legislation for these areas were rejected in the 1980s. The existing reserves have not received significant resources or special attention (Nair, 1993). The Chipko experience resulted in the formation of the Nanda Devi Biosphere Reserve which local people view with disdain based on the similar structure and anti-people practices relative to other national parks and sanctuaries (Mitra, 1993). There seems little hope for this concept as long as its implementation involves the same management procedures by the same institutions under the same legal structure which has produced the existing problems.

A second participatory approach which has received a higher profile in protected areas has been "ecodevelopment". Ecodevelopment refers to broad rural development initiatives for local communities which exist on the periphery of protected areas. The assumption is that if local lands and communities outside the protected area are made more productive and economically viable in their own right, then people will reduce their dependence on resources within the protected area (Indian Institute of Public Administration, 1994; Saxena, 1995; Singh, 1994). This concept has been initiated and promoted at international and national level through the Global Environment Facility (GEF), the Ministry of Environment and Forests, and the Indian Institute of Public Administration. The Government of India is initiating a US $67 million (Rs. 294.9 crore) project to cover seven national parks with two-thirds of this funding coming from the GEF. There are hopes to expand it to 40 areas with additional international funds. Although the proposal covers some of the past structural issues which have produced the alienation of local people, it still embraces the policy of eliminating local people's traditional rights and activities within protected areas (Indian Institute of Public Administration, 1994). It focuses instead on giving local people alternative activities, and educating them as to why conservation is important and why it is in their own interest to protect the areas. The existing park authorities are initiating the process through a top-down planning process where local people's views are solicited by a professional research team, but at no point are they given a decision-making or planning role in defining the approach.

The project has just been approved but is fraught with major deficiencies. There are numerous powerful reasons to reject this approach:

- There is little evidence to support the assumption that increasing the incomes and resources of local people will result in the reduction of their use of forest resources.

- The assumption that local people can be educated to see that conservation and elimination of their rights in protected areas is in their own interest when they lose critical resources in the process is unwarranted.

- The approach assumes in line with existing legislation that people are not living in parks, and thus makes no provision for the people who are, implicitly suggesting that they should be evicted.

- The assumption that large amounts of funds at international and national level will actually reach local communities through government structures is unwarranted, given the widespread consensus that this has not occurred through rural development programs. The funds disappear through administrative overhead and corruption during the process, leaving only the bureaucrats and policy-makers better off.

- The assumption that there can be meaningful local participation and decision-making, when the project is driven by enormous levels of funding at the international and national level, is invalid.

- The approach is not economically viable, in that it demands enormous resources for small areas, although most of the problems cover enormous areas.

We would argue in fact, that this approach is extremely destructive, in that it amounts to providing funds to the existing bureaucratic and policy making elites in the name of and, at the expense of the rights and interests of local people and the environment. It represents the newest example of how international and national elites define programs that appear to articulate "progressive" attitudes but in fact simply reinforce the roles of those currently in power. In rejecting this approach, we are not rejecting rural development for its own sake which is implemented through a grassroots community development process. But that is not the essence of ecodevelopment.

Joint Protected Area Management

A second group of alternative approaches has been proposed under the umbrella of "joint protected area management" (Chalawadi, 1995; Kothari et. al., 1995; Ramprakash, 1995; Sarabhai et. al., 1991). They acknowledge the failure of the past anti-people approach to protected area management. Kothari et. al. (1995) defines the new approach as:

> Joint Protected Area Management (JPAM) is the conceptualisation, planning, and management of protected areas and their surrounds, with the objective of conserving natural ecosystems and their wildlife, while ensuring the livelihood security of local traditional communities, through mechanisms which ensure a partnership between these communities, government agencies, and other concerned parties. (p. 2763)

The authors note that this definition shrouds a range of complex issues in specific settings, such as what should be the objectives for conservation and security of local livelihood, what should be the responsibilities of the partners, what should be the structures and mechanisms for implementation. They suggest that the approach will need to review the existing protected area network in relation to JPAM potential, empower local communities and engender a sense of ownership in the protected areas. It will also have to press for overall changes in the wildlife laws, and provide clear definitions of traditional rights.

The only specific proposal for JPAM relates to Shoolpaneshwar Wildlife Sanctuary in Gujarat (Sarabhai et. al., 1991). It is not appropriate to evaluate or draw conclusions about a model based on one example, but it deserves careful examination as it is the most explicit proposal to date for joint protected area management. It solicits input from leading proponents of the general JPAM approach. The report is to be commended for documenting its methods in such a way that the process and recommendations of the approach can be examined. Although this example speaks to the philosophy and concepts of JPAM, the process of implementation and the nature of the recommendations do not operate them.

(a). Project Process: This study of the park/people issues in the Shoolpaneshwar Sanctuary was undertaken by VIKSAT, a policy level NGO under the Nehru Foundation for Development which has had significant experience in joint forest management initiatives outside protected areas. It is unclear who requested or initiated the study, but it was not requested by the local people. The research was funded by the World Bank and was undertaken by a VIKSAT team of researchers who consulted a wide range of national policy experts, in addition to local forest officials, and local people in the area.

A detailed examination of the project process shows little evidence of it embodying the community participation orientation it espouses. There was no villager concerned in the planning or implementation of the research strategy, and there is no evidence to suggest that villagers were consulted about the report's conclusions. There are 77 villages in the sanctuary, yet the research team visited only four in approximately eight to ten days. Three of the four villages were visited in the company of a forest official, while in the other instance the researchers were unaccompanied. From the report, it appears that the visits were restricted to day-time meetings (approximately two per village) including travel time to these remote areas. Villager views on their present problems were solicited through participatory rural appraisal (PRA) techniques, and their views are interspersed through the report with data collected from a range of sources. In two of the villages the researchers report being viewed with suspicion and hostility.

There appear to be other villager perspectives not represented in the study. One report indicates that "tribals have been demanding denotification of Shoolpaneshwar Sanctuary, where forest officials were harassing the villagers for collecting traditional forest produce, while simultaneously allowing a paper mill to harvest large amounts of bamboo" (Kothari et. al., 1995, p. 2755). The research team did not find proponents of this point of view. On the other hand, among their policy consultations, they spent considerable time "brainstorming with" and "receiving suggestions from" (p. 1) experts from the Wildlife Institute of India, who have been among the lead proponents of evicting the Van Gujjars from the proposed Rajaji Park. Finally, the report notes that there are local NGOs in the area in touch with the people, yet there is no evidence that the team contacted them or solicited their views. NGO involvement is proposed in the joint management process. In short, this proposal process did not adopt the partnership approach for the involvement of local people that JPAM espouses.

(b). Project Content: The study recommends a mixture of joint forest management and ecodevelopment approaches. Ecodevelopment recommendations include reducing agricultural land and domestic cattle while developing more efficient production systems, promoting rural development efforts, and involving local people in alternative tourism initiatives. The approach goes beyond ecodevelopment in recommending that the villages remain within the sanctuary, continue to collect minor forest products, and participate in joint management committees. Yet most of the space is spent on justifying the joint management approaches in principle rather than describing how they might be implemented. They state that for the villagers "the concept of the sanctuary is new in itself, if known at all"(p. 90). They note that senior forest officials would have to play "an especially important role" in the process (p. 87). Joint management is expected to require "a very different approach to be adopted by the Forest Department," including their "sharing of (a) certain amount of authority with the local communities who are traditionally viewed as the main culprits in degrading the forests" (p. 87).

The researchers recommend training as the main process to bring about this change in perspective among forest officials, and make no proposals for structural changes in the park administration beyond "involving local people in the process." It is extremely difficult to envisage, given their description of the context and power discrepancies, how forest officials could sit on the same committees with the local tribal population and share management in an equitable fashion. The study recommends much more study (which would inevitably require more funding) before developing a concrete management plan, requiring two to five years and assessing every village in the sanctuary. It is unclear how the existing problems would be addressed during this time or how long it might take to implement a new approach after further study is completed. The suggestions for further study imply similar structures and processes to this study with no indication that local people would play a larger role.

In brief, although not recommending anti-people approaches as in the official ecodevelopment proposals, there is little evidence in either the process or recommendations of the Shoolpaneshwar study to suggest that useful local involvement and responsibility would actually occur. The research on joint forest management indicates that this involvement is integral to success. Since the participation proposals are vague, one cannot make a final judgement. The study points out however, that a major revision and reformulation of the existing Forest Department approach is required, yet beyond the provision of training opportunities by outside consultants and forest officials, there are no mechanisms to ensure that this would occur. Given the need for developing and testing alternatives in the present context, there may be much to learn from implementing this proposal. Possibly these critiques would be proven wrong, and/or possibly the approach would be revised. Moreover, the general discussion and debate about JPAM may mobilise some people to consider change within the Forest Department.

Given that there is only the one explicit proposal for the JPAM approach in India, it is also important to review JPAM from the much broader base of JFM experience outside of protected areas. As noted previously, the ecological and people problems are similar whether one is in a protected area or a conventional forest except for differences in law, the more prominent wildlife purpose of protected areas, and the level of Forest Department presence.

The assessment of JFM experiences outside protected areas concluded that emphasis should be placed on the community management end of the continuum, whilst recognising that innovative forest officials may be valuable instigators of change in areas lacking in community initiative and strength.

JFM has had most success where individual officers had the flexibility to initiate change at a local level. The likelihood of an officer having the freedom to innovate however, is even less within protected areas because of the higher Forest Department presence and focus. In addition, given the even higher level of mistrust and conflict between protected areas authorities and local peoples, and the far greater level of authority that forest officials have over local people, it seems naive to argue that forest officials can jointly share power and management responsibilities without structural change in their power relationships.

The goodwill and training approaches of joint forest management seem insufficient for the task on their own. There is far less reason to believe that joint forest management approaches will actually be joint or that a high level of Forest Department control at the outset will evolve into community responsibility. If not, there will not be enough return for local people, and they will not have the required stake in the process to make it successful. Though we present these philosophical and practical reservations with respect to JPAM, we hope they may be proven wrong as there is a desperate need for new approaches, and one model will not fit all situations. It would seem that the best context to implement JPAM is one in which it is requested by highly motivated individual forest officials in contexts which have had relatively lower levels of Forest Department - community conflict and mistrust. Regardless of the judgement on the viability of JPAM in general, the Rajaji context lacks the fundamental characteristics which would make this model even plausible here. The Uttar Pradesh government took more than five years to draft its Joint Forest Management Guidelines after the original 1990 central government circular requesting them. Even then, they were only drafted in response to external pressure(Saxena, 1995b). Although JFM was initiated in some states prior to 1990 and in others shortly after, there has been no action in Uttar Pradesh. The guidelines are also inappropriate for the Rajaji context, in that they do not apply to groups living within protected areas. They must leave the protected area, as only then is it possible for them to share in some of the benefits of the fringe zones. From the Van Gujjars' perspective, this amounts to the same eviction approach which has proved unworkable and exploitive over the past decade. Saxena et. al. (1995b) also note that the guidelines are weak, in that they allow for broad groups of people to claim rights over given forest areas, regardless of proximity and present usage, thus failing to implement the key community forest management principle that discrete communities should manage discrete areas based on present usage. The general problems identified in JPAM, however, evident in the Shoolpaneshwar proposal, suggest that there is a critical need to develop an alternative conceptual model which would extrapolate the experience in community forest management to the management of protected areas.

Community Forest Management in Protected Areas (CFM-PA) as an Alternative Approach

There is extensive evidence to indicate that the present system is failing, and participatory management approaches such as ecodevelopment amount to the present system with a modification in style but not substance. JPAM is untested but there are substantial reasons to question its potential efficiency. There is clearly a need for additional alternatives, but what reasons are there to suggest that a community-based approach is the best strategy in a positive sense?

Firstly, the conclusions from the research on community management experiences outside protected areas indicate its effectiveness, when communities have the specific characteristics correlated with success, or are capable of developing them. It makes sense to evaluate CFM-PA initially in conducive settings, and then examine flexibility across other contexts. From a resource perspective, it is cheaper than other alternative because it is based on communities taking over existing Forest Department roles. The relatively low cost avoids subservience to large international donors who in turn have specific vested interests. Placing power in the hands of local people with monitoring by government officials also provides an obstacle against vested government and political interests overwhelming the reform process. With their existing investment in protected areas, forest officials will inevitably fulfil their monitoring role.

Community management is inherently more respectful of the rights of indigenous people, providing mechanisms through which traditional knowledge is given credibility and validity in the wider society. Finally, if successful, the process itself can develop the capacity of the community which builds cultural pride and identity that in turn can be transferred to other contexts. It could provide a practical example of a viable alternative lifestyle to the dominant urban consumer mentality that is sweeping the world and is a major element of environmental destruction. Protective area management might become an example of the positive environmental practices of alternative cultures, rather than a means to distract attention from the broader environmental and social problems of mainstream culture. It may have much to teach mainstream society about environmental management.

If Community Forest Management in Protected Areas (CFM-PA) is to achieve the above aspirations, the model must be constructed based on the breadth of available knowledge, and must be applied through specific realistic proposals which can be implemented in a practical and effective manner. The primary purpose of the preceding section has been to synthesize available knowledge in order to identify important attributes for the approach. We define community forest management in protected areas as:

> Community Forest Management in Protected Areas (CFM-PA) is a set of organisational structures and processes for defining and managing protected areas in which the local peoples who have traditional rights to inhabit and/or use the area become the leaders in managing the resource, while government departments and other stakeholders have a monitoring and support role. The overall goal is to protect and administer, in a sustainable manner, the ecosystem, its wildlife, and the traditional rights and lifestyles of the local people integral to it.

A full consideration of the model, its principles, and the structures for implementing it in the Rajaji area are elaborated in Part IV of this report, after a thorough review of the ecosystem and local people's issues in the proposed Rajaji National Park. This report represents the first detailed proposal for community forest management in protected areas in India.

Community Management Approaches in Protected Areas in International Settings

Before proceeding to examine the history and management of the Rajaji context and to define the alternative CFM-PA proposals, it is important to recognise that this sort of approach has gained prominence recently in a number of countries across the world. In Nepal, the tradition of village forest guards which existed before government nationalisation of the forests in 1957 has been reintroduced in Sagarmatha National Park. It is used in every village in the daily management of the park, though as yet does not have legal standing. Local people have been trained in park management, and have occupied higher level positions in the park structure, including that of park director (Sherpa, 1993).

The Kuna Indians in Panama, when confronted with land encroachments from outsiders, created an organisation to protect and manage their forest areas. Working with outside experts, they mapped the area, established boundaries and began to manage the area as a biosphere reserve. A decade of management has provided protection for the people and the area, though now there is increasing local population and increasing pressure from the government to eliminate protected area status to facilitate development. The Kuna have responded with a management plan to reduce their pressure on the area. They face financial problems however, in implementing the plan, as they do not have government funding (Archibold & Davey, 1993).

In Northern Canada, a new national park was recently declared on Inuit lands, recognising all their traditional hunting rights, giving them the power to screen archaeological research and involving them in the drafting of the park's guidelines and management plans (Cordell, 1993). In Venezuela, a biosphere reserve and national park was recently declared in a tribal area and the management committee is a multi-agency group including the tribal people (Centeno &

Elliot, 1993). In both New Zealand and the Philippines, governments have worked out communal land leases with indigenous groups, whereby these groups manage their areas while keeping to specific land use covenants (Cornista & Escueta, 1989; Gould & Lees, 1993). In the Philippines example, the lease initiated in 1974 includes the mixed use of 14,000 acres of forest lands and six villages. The management structure relies on the traditional tribal decision-making system (the "tungtungan" of the Ikalahan people) for planning and dispute resolution. It includes decision-making through consensus with disputes being resolved by a council of elders, not dissimilar to the Van Gujjar panchayat to be discussed in part III of this report. The process has proved successful in terms of both rural development and environmental protection, though the indigenous organisation has few funds and there have been jurisdictional disputes with the local civil authorities (Cornista & Escueta, 1989).

According to Jeffrey McNeeley, Secretary General, Fourth World Conference on National Parks and Protected Areas, 1993, the management of protected areas should....

- Build on the foundations of local culture.
- Give responsibility to local people.
- Consider returning ownership of at least some of the protected area to indigenous people.
- Hire local people.
- Link government development programs with protected areas.
- Give priority to small scale local development.
- Involve local people in preparing management plans.
- Have the courage to enforce restrictions (with enforcement by local people).
- Build conservation into new national cultures.
- Support diversity as a value.

Finally, in Indonesia, a nature reserve was established in 1988 on sixty-eight square kilometres of land belonging to the Hatam people, after extensive consultation with them and their approval of the management plan (Mandosir & Stark, 1993). They participated in defining the reserve area which was sub-divided into sixteen management sub-areas based on collective groups of landowners willing to work together. As in other locales there were initial difficulties between villages so they were further sub-divided in a couple of instances to create smaller groupings. The Hatam agreed to maintain the resources within the reserve, and yet still kept their rights to hunt and use the forest in traditional ways. Their management committees use government sanctioned powers to enforce the regulations of reserve within their area, both with respect to their own members and outsiders. The committees are also involved in alternative economic development projects; and both the reserve and the Hatam appear to be prospering.

The range of recent experiments appears promising, as indigenous people begin to take responsibility for managing protected areas. It is noteworthy that the most sophisticated projects involve homogeneous tribal cultures. The experiments where they have been given the highest degree of management responsibility tend to be those in which the government already recognised the land rights of the local people such that the tribal group was negotiating powers from a position of strength. Hopefully, successes in these areas will encourage governments to grant local people more management autonomy regardless of official land title. Dasman (1988) argues that there is a fundamental distinction between "ecosystem people" (indigenous subsistence peoples) who depend on the ecosystem in which they live for their survival and "biosphere people" who depend on the whole biosphere and simply shift to another ecosystem if one deteriorates. Destruction of the ecosystem destroys ecosystem people, but is only a small problem for biosphere people. He argues that biosphere people create national parks, while ecosystem people have always treated their ecosystems as homes which they must care for and protect over generations.

J. McNeeley (1993, pp. 253-257) summarised a compilation of a wide range of initiatives with respect to indigenous peoples and protected areas across the world, with recommendations that are embodied in a community approach. CFM-PA as it is proposed for the Rajaji area incorporates these principles, based on international experiments in protected areas, and community management structures and characteristics that have demonstrated success outside protected areas in India. It is now appropriate to examine the local Rajaji context in detail, before defining the specific model and structures.

APPROACHES AT A GLANCE

Joint Forest Management	*Community Forest Management*
1. Based on existing system (Colonial system)	Based on traditional system of management.
2. Requires no structural change in power relationships, i.e. between forest department and local people.	Power rests with the people.
3. Less stakes involved for people participating and therefore less responsibility.	Full stakes involved and so a greater sense of responsibility.
4. **Attitudinal change** - partially successful only when innovative forest officials are involved.	**Structural change** - People take initiatives.
5. Cannot be replicated as such in a Protected Area where forest department - local people relationships are soured.	Applicable to all kinds of forests.
6. Problems in implementation once joint forest management gets institutionalized.	No problems in implementation as each plan is case specific and made by the people.
7. Lots of money involved for infrastructure, management and monitoring - dependence on large international donors.	It is cost-effective with no dependency on large international donors.

PART- TWO

THE RAJAJI ECOLOGICAL CONTEXT AND PROBLEMS

Chapter 3
Research Methods

*"Man is a part of,
not apart from, nature."*

F. Fraser Darling

A clear description of the four types of research methods utilised in this study is important to allow the reader to interpret the subsequent findings. They were:

(1). Extensive field research, in all ranges in the proposed park area and periphery, in order to identify ecosystem status and problems. This work included extended visits and interviews with a wide range of Van Gujjars to gain their perspectives and wisdom with respect to the ecosystem, their lifestyle, and their specific proposals for community management.

(2). Field visits and interviews with a range of local villagers living on the periphery of the park area who depend on the park for grazing, fuelwood, and/or bhabbar grass.

(3). Interviews with forest officials and other important stakeholders about to their perspectives on ecosystem status, problems and directions for the future.

(4). Review of relevant written literature from local, national, and international sources.

Field Research Methods
Field research in the park area included visits to all forest ranges and nearly every block (see Appendix 1). The researchers explored the forest, both in the company of the local Van Gujjars and on their own. They observed and/or participated in all aspects of Van Gujjar daily activities. For example, lopping was observed on many occasions in numerous khols, including instances where the individual Van Gujjars were not aware of the researchers' presence. Transit between khols

provided another opportunity for detailed observations of the local conditions. There were extensive discussions in the deras with the Van Gujjars, and researchers used time in the forest with individuals as an additional opportunity to ask questions and gain information about the forest and their lifestyle. Discussions in deras typically involved not only one particular family but also neighbours who were called to the discussions. There were also two extended trips to visit Van Gujjar deras in the mountains and, a number of discussions with Van Gujjars at stopping points on the autumn migration. Appendix 1 presents the dates and numbers of visits through the seven month period of field work. In the course of seven months the researchers visited 84 deras and talked with more than four hundred Van Gujjars.

The research team had extensive training and experience in a wide range of interview research techniques. Mr. Simronjit Singh served as an assistant/translator during his first ten weeks of fieldwork. During this period he was trained, and practised interviewing and note taking under the supervision of the senior researchers. He subsequently completed field visits and interviews independently. During the first five months of interviews, the researchers were usually introduced to a dera through an RLEK field worker or volunteer who was familiar [you can say "familiar with" as "known to" according to the required emphasis] with the family. Since RLEK staff have established long-term relationships with the Van Gujjars, based on the NGO's work and advocacy efforts over five years, the researchers were always welcomed as friends and colleagues. We fully described our backgrounds, our task, and our relationships with RLEK.

Interviews were multi-faceted involving a wide range of questions. They were conducted in an informal style according to the pace of our hosts, with initial emphasis on relationship building. The questions however, were based on a detailed list of topics and specific queries. These questions had in turn developed from knowledge gained through initial interviews and reading. Detailed field notes were recorded as soon as possible after the interviews, but not during them, so as to avoid making those interviewed feel uncomfortable. The same questions were asked to different Van Gujjars in order to ascertain the breadth of perspectives and to cross-check the veracity of specific responses. Question sequences were initiated with general brainstorming types of queries, and only followed-up with more specific questions after the person interviewed had exhausted his or her ideas on the general topic. We would challenge their ideas by posing specific problems they might encounter, e.g., "if Van Gujjars managed their own khols as a committee, what would stop more powerful Van Gujjars from intimidating less powerful ones?" The same information was sought through multiple questions in instances where we were concerned that there might be misunderstanding or reason for the person to be circumspect. Extensive time was spent cross-checking answers and looking for inconsistencies and discrepancies in patterns of responses. Field observations were used to challenge or verify information. We followed up with questions to RLEK staff or other Van Gujjars if particular responses could have been based on hidden agendas. As our knowledge base increased over time, we used it to push detailed lines of questions and problems.

This careful, detailed interview strategy verified our general feeling that Van Gujjars were very direct, open and honest in the vast majority of situations. They would frequently share information which could be seen as portraying them in a negative light (e.g., that there were too many Van Gujjars in a given khol in relation to the ability of the forest to regenerate) and were detailed and realistic in defining their problems and those of the forest and wildlife. Even later in the research when draft ideas had been defined, we would ask each person for their own ideas at the outset and only then share the ideas synthesised from previous discussions. We always encouraged them to follow-up by discussing the topics with others, including criticisms and problems. We carefully explained that the ideas would be synthesised into an alternative community management plan which would be presented to the government, and that the plan could not be implemented without government approval. We clearly stated that we could make no promises other than to synthesize and write down the ideas for consideration by all interested parties.

Interviews with Local Villagers

A second major element of the field research was visits to local villages around the and detailed interviews with the villagers (see Appendix 1 for list). Our first visit to villages on the southern park boundary was facilitated by VIKALP, a local NGO which has supported the local people in their conflicts with the Forest Department and their demands for development programs.

We toured villages along all portions of the park boundary, but especially focused on talking to villagers in the pilot area, visiting every village in the area in order to be able to define their specific proposals for this area. The village visits in the pilot area were arranged by a well-respected Pradhan of one of the larger Taungya villages, and we were warmly welcomed in all instances. We clearly explained our purpose again in terms of putting proposals together for community management of the proposed park. The interview process followed the same format and approach described above in relation to work with the Van Gujjars. We also visited Rasulpur village on the edge of Luni block in the Chilla range, and a settlement of Tehri Dam evacuees in the elephant corridor. Finally, we visited two villages in the mountain area adjacent to Van Gujjar summer pastures near Mori in order to gain a village perspective on interaction between the villagers and the Van Gujjars in the summer.

Interviews with Forest Officials and Other Stakeholders

We particularly solicited views from a range of forest officials in the Rajaji area, from the Conservator for the region to local forest guards. We also met with a number of NGOs locally, including VIKALP, Friends of Doon, Himalayan Action Research Centre, and researchers at the Wildlife Institute of India. We consulted NGO staff in Delhi at the Centre for Science and the Environment, Society for Promotion of Wastelands Development, World Wildlife Fund-India, and the Centre for Environmental Law.

Review of Relevant Written Literature

We obtained a wide range of written material through RLEK, the Forest Department, the Wildlife Institute of India, the Forest Research Institute, and other resource libraries. We have relied on the wealth of empirical data in the recent official proposed park plan (Kumar, 1995).

National Workshop and Development of the Final Report

In synthesizing the research findings, different perspectives and data were compared in order to identify areas of agreement between the various constituencies and areas where there was disagreement. Areas of disagreement were then further analysed in relation to empirical research data and our direct field observations in order to draw conclusions.

A preliminary draft of this report was then prepared for discussion and consultation purposes at a national workshop in Dehra Dun on February 24 and 25, 1996 at the Indian Institute of Petroleum. The opening inaugural session included approximately 500 Van Gujjars plus the workshop participants and invited guests. More than sixty people participated in the two days of technical sessions including Van Gujjars, local villagers, Forest Department officials, social activists, legal scholars, government officials, policy-makers, members of the press, and NGO representatives (see Appendix 2 for list of participants and workshop agenda). This diverse group of people discussed the draft proposals, and made recommendations for revision which were incorporated into the final report. Their contributions were invaluable in framing the final document.

Chapter 4
The Ecological Context

"Protect environment for the people, rather than from the people."

International Project on Community-Based Law Reform Proposals
Faculty of Law, University of Delhi, 1995

Chapter 1 of this report identified the international and national policy which has compounded the problems of the proposed Rajaji National Park. Chapter 2 examined potential policy alternatives to address the problems based on experience in other jurisdictions. Each locale however, is different and it is critical to provide a detailed description of every one. Only then is it possible to analyse local issues and present local strategies to resolve them within the context of national and international lessons. It is essential to take a step by step approach based on data and sequential logic in order to define major problems and their contributing factors. This section is sub-divided as follows:

- Description of (1) the environment, (2) local people, and (3) government and commercial land uses.
- Data is then reviewed on present ecological status and trends, in relation to the impact of human activities. The nature and magnitude of the ecological problems are identified, followed by a careful analysis of the factors which seem to have caused these problems.

Description of the Environment

Description of Area

The proposed Rajaji Park comprises 825 square kilometres of land situated in the Shiwalik hills, and is representative of the Shiwalik ecosystem which lies between the Himalayas and the Upper Gangetic plains spanning a 2100 kilometre band in India and Nepal. The proposed park represents the integration of the Chilla, Rajaji, and Motichur Wildlife Sanctuaries and an additional section of reserve forest (see Map 2). Notification of intent to create of the park was made in 1983, but local claims and rights have not as yet been settled as per the Wildlife Act, hence final notification for the park has not been issued.

The cities of Dehra Dun, Rishikesh, and Haridwar are situated at three corners of the proposed park with the small market town of Mohand on the Southwest corner. The Dehra Dun District on the Northern boundary of the proposed park included ten lakh people in the 1991 census, the majority living within twenty kilometres of the designated area. The Ganges river bisects the land as does the major railway line from Delhi to Dehra Dun, and a major road from Haridwar to Rishikesh and Dehra Dun. There is a railway station at Kansrao in the core area of the proposed park.

In addition to the proposed park area, this study also includes the forest to the west of the park in the Mohand area of the Shiwalik Forest Division. This area is studied because it is ecologically contiguous with the western park area, and includes a major population of Van Gujjars interconnected with groups within the park area. The official park plan is proposing to include this area in the buffer zone of the main park. The road from Delhi to Dehra Dun divides the proposed park from the Mohand range. The total area includes eight forest ranges in the actual park area and two in the Shiwalik Forest Division (see Map 2). These ranges are further sub-divided into forest blocks, which are periodically mentioned through out the report (see Map 3).

This area has been suggested for a national park, as it is among the better preserved remnants of the Shiwalik ecosystem. It represents the Northwest limits of the tiger and elephant populations in India, and includes a wide diversity of flora and fauna.

Climate

The climate is similar to that of the northern plains but is moderated by the Himalayan hills and somewhat higher elevations. There are three seasons. Winter is from November to

February with modest daytime temperatures, cool nights with dew and frost, and a little of winter rainfall. Dew fall diminishes through March and April as temperatures rise to summer levels in May and June. Temperatures rise as high as the mid 40s, and the land becomes parched and dusty with an occasional shower or thunderstorm. The hot season breaks in late June with the onset of the monsoons. Pervasive humidity with heavy rainfall continues throughout September.

Terrain

The terrain can be sub-divided into the northern and eastern slopes of the Shiwalik Hills which fall into the Doon Valley and the Ganges River basin, the southern slopes of the Shiwaliks dropping to the Gangetic Plains, and the outer Himalayan hills and bhabbar tracts on the eastern bank of the Ganges which comprise the Chilla Sanctuary.

The northern and eastern Shiwalik slopes are steep in the upper reaches, but flatten out lower down. They are cut by a series of "raos" (wide stream or river beds, largely dry except in the monsoons) with ridges running along them. The raos in this region generally flow into the Suswa and Song rivers which drain these slopes and are perennial sources of water. The slopes usually suffer from lack of water for most of the year, and are a mixture of sandstone and interbedded clay with loose gravel in the Doon Valley and raos. Soil cover is thin, non-existent in the raos. Soil erosion is severe as most rainwater drains off the slopes quickly or percolates through the thin soil and loose sandstone. This results in a fragile ecosystem which has its productivity constrained by water shortage for most of the year while being ravaged by erosion and superabundance of water during the monsoon season. These factors also slow down regeneration from major cutting and clearing.

The southern slopes of the Shiwaliks also have varying blends of loose sandstone and interbedded clays which are segmented by raos and ridges, which ultimately drain into the plains and the Ganges River. These slopes tend to be more rugged than the northern and eastern reaches. Water availability is again a problem in the dry season on these slopes, although soil is somewhat deeper in lower gradient areas.

The outer Himalayan hills to the east of the Ganges include slopes which can vary from steep to gradual with parallel sets of north-south valleys grooved with them. Many areas of these hills are steep and difficult to access. The hills terminate in the flat plains' forests of the proposed park adjacent to the Ganges.

Forest Types and Habitats The vegetation of the Rajaji area is dominated by sal forest including several sub types: Moist Shiwalik Sal, Moist Bhabbar Dun Sal, and Dry Shiwalik Sal. These cover 75% of the proposed park. Other forest types represented include West Gangetic Moist and Northern Dry Mixed Deciduous Forest, Low Alluvial Savannah Woodland, Khair Sissoo Forest, Sub-Montane Hill Valley, Dry Deciduous Scrub, and Sub-Tropical Chir Pine Forest. Aerial photo interpretation of crown density for 1981, photos covering approximately 75% of the park area, demonstrate the vast predominance of forest land made up of sal or mixed forests (see Table 2). Approximately half the forest has more than 60% crown cover, a third of the area has 20%-60% crown cover, and 16% of the forest has less than 20% crown cover. The photos identify only 1100 acres of active plantations (2% of the surveyed area). It is difficult however, to identify plantations from the photos and land use statistics for the proposed park indicate 7.8% of the area is under plantation. There are significant differences in the quality of the forest as one moves across different areas of the terrain depending on gradient, soil, climate, and water conditions, human interference, and past forest practices.

There are few plant species which are restricted only to this region, but three species are endangered: Erenostachys Superba, Cataniecis Bracha, and Euphorbia Candicifolia. Swampy patches which tend to have deeper soil are scattered through the area, and have particular ecological importance in terms of biodiversity and as wildlife habitat during the dry season when water is scarce in most places.

The availability of vegetation as food for animals depends on the season. The monsoon promotes rapid and vigorous regeneration of all grasses and plants, and water is widely available. Herbivores have a wide choice of habitat and ample food supplies. There is continued availability of food throughout the winter, based on growth during the monsoon and the re-charging of greenery through winter rains. Water availability decreases through the winter, but is still easily accessible. As the temperatures rise and the ground hardens with the advent of summer, water holes diminish and grasses and ground plants dry up. Food includes the flowers and fruit of woody plants, fallen leaves, sal flowers which fall to the ground, and the new seedlings which emerge. The summer is a season of scarcity for wildlife, and their ranges are more restricted then.

Table 2
Aerial Photo Analysis of Proposed Rajaji National Park
(Density Classes of Forest In Hectares)

A. Forest Lands	60%+	20-50%	5-19%	Total
1. Sal	15,086	1,346	161	16,593
2. Misc. with Sal	3,963	2,890	246	7,099
3. Misc. Forest	8,738	14,414	8,601	31,753
Total	27,787	18,650	9,008	55,445
4. Forest Blank				345
5. Scrub				
(i) on gentle slopes				128
(ii) on steep slopes				103
6. Re-growth				67
7. Plantation				1,121
Total Forests				57,209
B. Non-Forest				
1. Agriculture & Habitation				738
2. Water Bodies				3.850
3. Grassy/Barren Waste Land				335
Total of Non-Forest Land				4,923
Total of Forest and Non-Forest Land		63,132		
Photo Gap				20,464
Grand Proposed Park Total				**82,596**

Excerpted from Kumar (1995)- Management Plan of Rajaji National Park, Vol. I, p.33

Animals

The elephant is considered to be the most important animal species for conservation purposes in the park area with an estimate of 522 in the 1993 wildlife census. Other significant large herbivores include the chital, sambar, barking deer, antelope, goral, and wild boar. Tigers and leopards are the high level carnivores though there are very few tigers and rarely seen. The census estimate of 15 to 20 may well be high. Leopards are more common. Lessor carnivores include the jackal, jungle cat, leopard cat, marten, and civet. Sightings of Rhesus macaque and the common langur monkeys are frequent in the proposed park. There is a range of reptiles including the king cobra, python, and monitor lizard. There are more than 315 bird species in the park area and their ranges extend far beyond this locale.

Description of Local People

Van Gujjar Population and Livestock

The Van Gujjars are nomadic forest dwellers who earn their livelihood by raising buffalo and selling milk in villages, towns, and cities through one or more of several marketing systems. They utilise the forest for fodder for their animals. Williams (1874, p. 29) notes that Van Gujjars came to the Rajaji area in the 18th century, where they have since lived for generations in "deras", (large thatched circular huts), throughout much of the park area and in the Shiwalik Forest Division to the west. Their deras are typically situated on the relatively flat edges of raos. They require a permit from the forest department for any particular location with respect to the use of the forest for animal fodder. The permits are based on the last official census, which dates back to 1937 and identifies 512 families. Given that the permits are outdated in relation to present population and distribution of families, several deras are often situated close to each other because they are sharing or have sub-divided one permitted area. There has also been widespread shifting of permits and Van Gujjars over the years by forest officials.

Researchers and officials have varied in their population estimates in recent years, but there have been wide discrepancies depending on the reason for the estimation. As a result, Rural Litigation Entitlement Kendra (RLEK) initiated a census of the number of Van Gujjars and Van Gujjar animals in the area of the proposed Rajaji Park in 1992-1993 as a basis for its work with the Van Gujjars. Census accuracy is difficult to achieve, requiring a careful process as Van Gujjars are circumspect about disclosing their populations. Iinformation they have provided previously, has been manipulated and used against them in the government's past attempts to remove them from the proposed park area. As a result, it was essential for RLEK to build constructive relationships to overcome their suspicions before initiating a survey. The detailed survey method is described in Appendix III.

Table 3 presents the population of human beings by range, and Map 4 demonstrates the distribution of deras. There are 5532 Van Gujjars living in the park area, and 1536 in the Mohand area. There are wide discrepancies in the number of Van Gujjars by khol and range. Dera location and population are dependent on forest department permits, and are in turn associated with fodder and water availability, terrain, and specific Forest Department interests and practices. Thus as there is a deficiency of fodder trees in the Motichur and Kansrao ranges, and few Van Gujjars. But high numbers of Van Gujjars live in some khols in the Dholkhand Range as this area has relatively good forest. Some khols in this range have been closed to Van Gujjars for a considerable period (e.g., Dholkhand) and a couple more recently (e.g., Malowala, Andheri). The Van Gujjars in these closed areas were moved to other khols, which has caused overpopulation in these khols (e.g., Gholna Khol). This trend of shifting Van Gujjars because of loss of their existing lands has been going on during the past century. For example, the working plans for the Chilla Range in the 1960s refer to problems created by clear felling areas for plantations, and then shifting Van Gujjars to areas which already had resident populations. Thus, Van Gujjars used to be spread over a far greater area, but Forest Department policies and associated urban and agricultural land encroachments have caused an increase in their population within the proposed park areas.

Table 3
Populations by Range of Van Gujjars and Van Gujjar Buffalo

Range	People	Buffalo
Ramgarh	439	1214
Motichur	332	826
Kansrao	298	714
Gohri	830	1206
Chilla	669	1135
Dholkhand	956	3034
Chillawali	1249	2588
Ranipur	759	1561
Proposed Park Total	**5532**	**12278**
Mohand Area*	**1536**	**3181**

* The Mohand area figures are for blocks from Mohand to Sahasra, larger than the present of the reach Mohand Range which now stops at the Kaluwala Block.

Estimates of the number of Van Gujjar buffalo and other animals has been a controversial topic over the years; thus, great care was taken with the census (see Appendix III), relying on actual counting and reports of nearby Van Gujjars rather than on self-reports. According to this census, there are nearly 12,300 buffalo belonging to Van Gujjars in the proposed park area with an additional 3181 in the Mohand area. There are also some cows and goats though far less than buffaloes. These figures represent actual buffaloes rather than official numbers as there is a widespread practice of allowing livestock above permit specification in return for fees, paid either in milk products or rupees. The number of buffaloes is clearly correlated with the number of people, the pattern of permits, and the availability of fodder and water.

Table 4 describes the size distribution of buffaloes herds by families (these are predominantly joint families with some nuclear families depending on how herds are grouped). There is a range from 1 to over 100 buffaloes per herd but the vast majority of Van Gujjars have between 1 and 30 buffaloes, with few families having more than 60 buffaloes. The number of buffaloes is a direct measure of the economic status of a family as most Van Gujjars are dependent on selling milk for their livelihood. It is clear from these statistics that there are minimal economic discrepancies between families in an absolute sense, and even "well to do" families have relatively few economic resources (see chapter 7).

A final major Van Gujjar population issue is the percentage of Van Gujjars and animals which is nomadic and moves out of the park area in the summer and monsoon seasons. Unfortunately, the RLEK census could not get a reliable estimate, with the Van Gujjars tendency to over-report that they are no longer migrating. Given their insecure status in the proposed park, they fear that they will lose their areas if they do not define themselves as settled. In addition, most families leave a few members behind in order to protect their deras from being destroyed when they leave. A family will define itself as settled and yet some of its members and many of its buffaloes will go to the hills like nomads. There are also numbers of officially settled Van Gujjars outside the proposed park area who area still nomadic. Finally, others move out of the proposed park to other local forests and, therefore, are still nomadic although they do no visit mountain pastures.

Table 4
Each Family's Buffalo Herd Size By Range

Park Ranges	0-10	11 to 20	21 to 30	31 to 60	60+
Kansrao	9	12	11	6	0
Motichur	13	18	8	6	0
Ramgarh	17	20	8	7	1
Gohri	42	48	11	8	2
Chilla	45	57	14	5	4
Dholkhand	26	71	27	23	2
Chillawali	15	28	28	9	2
Ranipur	48	57	15	8	0
Mohand	22	50	47	25	3
Totals	237	361	169	97	14
Percent.	27%	41%	19%	11%	2%

Based on our fieldwork in the proposed park area over seven months, it is evident that there are varying proportions of nomadic versus settled families depending on the area. As one moves from the Mohand range east to Motichur and Ranipur, the proportion of nomads going to the mountains decreases with high percentages in the west and lower percentages in Ranipur and Motichur. Most of the Van Gujjars to the east of the Ganges in the Chilla and Gohri Ranges are nomadic, moving out of the proposed park ranges to higher region in the summer season. Berkmuller (1986) and Clark et. al. (1986) estimated that 50% of the resident Van Gujjar families were nomads; this may be a reasonable estimate of families, but not buffalo or family members. Families with higher numbers of buffalo are more likely to be nomadic as they require great quantities of fodder per dera in the summer season. Based on our fieldwork we would guess that in most instances someone remains behind, yet many more than 50% of the buffaloes migrate out of the proposed park each summer. The factors and trends over a period of time in migration percentage are discussed in chapters 5 and 6.

Taungya Population

The Taungya people were encouraged to settle in the proposed Rajaji Park area from the 1930s as a labour source to grow and tend tree plantations for the Forest Department. The Forest Department had recognised that its internal procedures for growing plantations were failing, and hoped this regime would produce more success. The Taungyas were allowed to plant crops in between the rows of trees during the first few years, tending and weeding both the trees and the crops. The Forest Department would then shift them to a new area to plant trees and the associated crops. In addition, the Taungyas were given small quantities of agricultural land (124.5 hectares for the three villages inside the proposed park). By the 1980s the Taungya plantation were stopped, and the people remained in their existing locations with very small parcels of land. Their villages have never received revenue status, and the residents do not have a title to their lands.

There are nine Taungya villages on the southern boundaries of the proposed park, three are within the proposed park, and one which was within the proposed park has been forcibly moved outside the area. Park authorities estimate that there are 137 families and approximately 1000 people in the three villages within the proposed park. Berkmuller (1986) however, reports a total of 310 families in a survey of the same three villages. Our research team visited one village (Haripur) where the residents estimate was 650 families and 3500 people, and the park statistics suggest 450 families and 2800 people. Taungyas hold small numbers of cattle on a subsistence basis, but these are combined with the cattle of other villagers in estimating population.

Gothiya Population and Livestock

Gothiyas are a distinct cultural group, formerly nomadic herders, who have used the Chilla area for generations for winter grazing lands. They settled in this area however, many years ago, and some were given a thirty year lease in 1975 which was later cancelled. The official park plan reports 47 families, 167 people, and 261 livestock. Some Gothiyas were evicted from the proposed park area in the 1980s and have settled on the periphery.

Tehri Dam Oustees Villagers

Park authorities report sixteen families residing in the proposed park area occupying 49 acres of land in the primary elephant migration corridor. They were settled there in the 1960s by the government due to the evacuation of the Tehri Dam area.

Villages Within the Proposed Park

There are nineteen revenue villages with approximately 2500 families reported within the proposed park boundary, predominantly near the Ganges river and to the east of it. Sixteen villages are in the Pauri Garwhal District and three in the Dehra Dun District. Berkmuller found that village estimates of population in 1986 outside the proposed park were far higher than 1981 census figures; thus, it seems likely the population is underestimated.

Villages on the Park Periphery

Villagers from a distance of up to 10 kilometres utilise the resources of the proposed park for grazing, fuelwood, bhabbar grass, and fodder, with heavier dependence by those closer to the boundary. The park statistics estimate that there are 105 villages with potential dependence comprising 1.65 lakh people. One would expect these figures to be significantly low in relation to the actual population as the source is the 1981 census data. Tables 5 & 6 provide statistical data for these villages for population, livestock, and land use by district.

Livestock is estimated at 77,032 animals. The figures for the Haridwar district, comprising 35 villages and 28,898 animals to the south of the proposed park, seem compatible with a survey by VIKALP which found that 21 of these villages had 20,551 animals (VIKALP, 1994). The proposed park permits 64,000 cattle to be grazed in the park areas throughout the three districts. There is very little forest resource outside the proposed park boundaries in the peripheral area. Only 11% of this peripheral area is forest and 60% is devoted to agriculture.

Table 5
**Numbers of People and Livestock In Villages Peripheral to the
Proposed Rajaji Park**

Number of people and livestock residing within 10 kilometres of the proposed park by district.

Name of District	No. Villages	Total Pop.	No. Livestock	Park Grazing Permits
Dehra Dun	34	78,482	31,410	12,000
Haridwar	35	63,094	28,898	30,000
Pauri Garwhal	36	24,354	16,724	22,000
Total	**105**	**1,65,930**	**77.032**	**64,000**

Table 6
Land Use of Area Peripheral to the Proposed Rajaji Park

Village lands in hectares, for those within 10 kilometres of the proposed park, by district.

Name of District	Forest Land	Irrigated Land	Non-irrigated Land	Waste Land	Unprod. Land	Total
Dehra Dun	985	4362	507	355	1597	7806
Haridwar	7	1731	9075	1650	4465	16,928
Pauri Garwhal	2453	1385	2374	617	761	7590
Total	**3445**	**7478**	**11,956**	**2623**	**6823**	**32,324**
%Land Use	**11%**	**23%**	**37%**	**8%**	**21%**	**100%**

Tables 5 & 6 excerpted from Kumar (1995)- *Management Plan of Rajaji NationalPark, Vol. I*, p. 177, and *Vol. II*, p. 241.

Description of Government and Commercial Land Uses

Proposed Park Facilities and Staff

The proposed park is administered by the Director at the rank of Divisional Forest Officer, and is allotted 255 staff positions. 223 of these positions were filled as of 1995. The proposed park is divided into eight ranges, twenty-four sections and fifty-eight beats. The ranger is responsible for one range, a forester for each section, and a guard for each beat. The park planners received reports that staff salaries are very low, housing is unsatisfactory, and working conditions in the forest are isolated as staff must live away from their families (p. 194). There are 207 buildings under park administration throughout the eight ranges, varying from rest houses to storage facilities and offices. There are fourteen vehicles for transportation. There are thirty-one sets of wireless transmission equipment for communication.

Other Government Department Land Usage

Seven other government departments occupy land within the proposed park amounting to 2183 hectares. 1700 hectares are devoted to power transmission lines and 373 hectares for the power generating project.

The 14 kilometre Kunnao-Chilla Power Canal was constructed on the east bank of the River Ganges in the mid-1970s, and provides a barrier to elephant migration across the proposed park from east to west. It is twenty-two meters wide with forty-five degree sloped cement banks which elephants can not surmount . The southern end of the canal houses a large power generation and transmission facility.

The army received 346 hectares of forest land in 1976 for a base, including sixty-three hectares within the proposed park for an ammunition dump which is critically located in what was the primary elephant corridor between the eastern and western portions of the proposed park.

The new Haridwar District Headquarters is presently being built immediately adjacent to the proposed park boundary in the Ranipur Range. Reserve forest land is being cleared to provide for the new buildings which are partially completed.

Transportation Corridors

The major rail route from Delhi to Dehra Dun passes through the proposed park between Haridwar and Raiwala, and again at Kansrao. There is a station within the core area of the proposed park. In addition, the major road from Haridwar to Raiwala and then to Rishikesh or Dehra Dun crosses the proposed park. The western boundary is the major road from Delhi, Saharanpur to Dehra Dun, which separates the proposed park from the Mohand Range of the Shiwalik Forest Division.

Industrial Uses

Major industrial facilities are located on the periphery of the proposed park, and have replaced previous forest land over several decades past. There is BHEL at Ranipur, IDPL near Rishikesh, squeal of wood-based industries at Jwalapur, Flex food industry and Birla Yamaha at Laltappar, and a group of industries at Kuanwalla.

Tourist Usage

Only the Chilla and Motichur Sanctuary areas are open to tourists from November 15 to June 15. including a number of wildlife-viewing towers. There are ten forest rest houses with 25 suites available for overnight stays. Between 1987-88 and 1992-93 the number of tourists per year steadily rose from 5558 to 10,485 people according to park statistics. In 1993-94, there was a drastic drop to 2880. 1992-93 was the year in which there was tremendous publicity regarding conflicts between the park authorities and the people living in and around it. This possibly, caused the drop in tourism during the subsequent year.

Temples

There are eight temples leased within the proposed park, and visitors are given the right of way through the proposed park to visit them. Several have been expanded in recent years. Six to seven lakhs of people visit the Mansadevi temple every year whilst approximately five lakhs of people visit the Neelkanth temple for the annual fair during the monsoon season. There are large numbers of people visiting these shrines year after year.

Chapter 5
Ecological Problems and Human Impact

It is essential to take great care in describing and analysing the ecological problems of the Rajaji area and the role humans play in creating them, in order to ensure that alternative management proposals address the problems appropriately. The official park plan paints a bleak picture of the ecological status of the proposed Rajaji Park (Kumar, 1995, vol. 2., pp. 1-4). It states that "habitat degradation is the major cause for concern in the proposed Rajaji National Park" (ibid., p. 63). It indicates that this degradation may be "judged by depletion in forest cover, and in ground vegetation, increased rate of soil erosion and weed infestation" (p. 63). In addition, it notes wildlife problems including water scarcity, competition with humans for food, restricted habitat due to encroachment, and habitat destruction due to fire. It is important to define the present situation/problems carefully, based on the available data, before proceeding to interpret. In accordance with the official park plan, we examine and discuss the following problems:

(1) Forest Deterioration
(2) Depletion in Ground Vegetation
(3) Weed Infestation
(4) Wildlife Problems
(5) Increased Rates of Soil Erosion
(6) Forest Fires

The *status* of each problem is examined followed by an *analysis of the contributing factors* as identified by park authorities and local people. Five major types of contributing factors have been identified: forest management practices and policies of the Forest Department, industrial and urban encroachment, illegal activities, Van Gujjar impact, and villager impact. Local people have emphasised the first three factors, while park officials have emphasised the latter two. Each ecological problem will be carefully examined with respect to these factors. In some instances there are complex relationships between factors which must be understood.

Forest Deterioration

Status

Park statistics present the types of forest in detail at the block level, but there is little information indicating density or quality of the forest and no comparison with previous data. In the course of our visits to every area of the proposed park, we consistently asked Van Gujjars to rate the forest as high quality, moderate or poor forest. Their ratings were based on their indigenous knowledge of the quality and density of trees, amount and type of vegetation, and soil and water conditions. Map 5 describes the Van Gujjars' reported patterns through all areas. In general, Dholkhand, Mohand, and Chillawali have high numbers of Van Gujjar buffalo and the better forest. There is no pattern of high numbers of Van Gujjar buffalo being associated with poor forest. Map 6 provides a similar set of ratings reported by the Van Gujjars for areas based on the quality of the forests *for* fodder. (It is possible to have a dense forest with little fodder).

The key issue, given the present forest quality, is to determine if there is a trend toward deterioration. Comparisons of crown density between 1981 and 1986 by different methods (see Table 7) indicate a slight, probably insignificant increase in crown density, but difference in methods would be the most likely explanation for this discrepancy. The only other means of determining a trend is to examine documented changes through the analysis of past management plans, and to ask people who have lived and worked in the proposed park for many years.

Broad view of the history of forest practices is described below, based on the perspectives of the forest officials who proposed and evaluated the working plans for the past century. The review clearly indicates that there has been significant deterioration during this period. This conclusion is supported by the opinions of a wide range of Van Gujjars we

---

interviewed who have lived in the area for many years: "The forests earlier were extremely dense, tall grass, and lots of wild animals, but it is no longer like that" (Gani from Mohand Range). Though the conclusion is not in doubt, however, it is very difficult to determine the speed of deterioration or the patterns through areas. On one hand there is evidence of significant degradation, but on the other hand the forests are still there, despite the one hundred years of intensive human exploitation.

Table 7

Forest Crown Density of the Proposed Rajaji National Park

Land Type	1981 Survey*	1986 Survey**
Dense forest (Crown Density > 40%)	610.5	620.7
Open Forest Crown Density 10-40%)	170.1	163.2
Water Bodies	28.3	28.3
Non-Forest	16.7	13.7
Total	**825.6**	**825.9**

* Remote Sensing Data
** Thematic Mapping of Aerial Photographs

Excerpted from Kumar (1995)- *Management Plan of Rajaji National Park, Vol. I,* p. 18.

Analysis of Contributing Factors

Each of the five major factors identified as potentially responsible for forest degradation are examined in turn:

(a) Forest Management Practices: Both park officials and local people identify forest management practices as a primary cause of forest deterioration. The park plan includes a careful review of the history of forest management over the past century in the area. This report will briefly synthesize it, using quotations from the official plan.

The conclusion of the report regarding to forest management and wildlife is:

> An overview of the history of management of the area clearly indicates that in the initial 60 to 70 years of management history, conservation and protection of wild animals was not the priority of forest managers. More emphasis was placed on the exploitation of forests and Shikar. Most of the prescriptions and creation of working circles in the management plan were aimed at more exploitation, better yield of forest produce, creation of more and more shooting blocks, and obtaining more revenue out of the forest wealth. Some of the presumptions like the removal of Mallotus trees, suppression of grasses, planting of eucalyptus in grassy blanks inside the forest area, were contrary to wildlife conservation and contributed in habitat destruction. Exploitation of timber and NFTP by forest contractors was another major cause of habitat destruction. (Kumar, 1995, vol. 1, p. 191)

In reviewing the plans of the 1960s and 1970s the report notes some change in philosophy, but in each case writes that conservation measures were either not proposed or not implemented. It states that, with the advent of the proposed park, regular efforts are being made for the protection of wildlife habitat.

It is clear that the forest management practices proposed by the Forest Department served the purpose of forest exploitation. In many cases, as noted by past working plans, the actual implementation of practices was even more destructive and exploitive than the measures which were proposed. Prior to the advent of the Forest Department in 1868, there was extensive exploitation of the forests by the British, which was followed by 100 years of deterioration. The park plan provides numerous examples of the destructive results of forest management practices for the Chilla area, after the Forest Department assumes authority in 1868 (see box below). The historical analyses of the results of forest management practices in the Motichur and Rajaji Sanctuaries make very similar statements and are not, therefore, presented.

Selected Results of Forest Management Practices in the Chilla Sanctuary

The following statements are direct quotes from the park plan (Kumar, 1995).

Hearle's Plan, 1888-1898

- *The sal forests did not contain many large trees due to past excessive fellings... (p. 75)*
- *The bamboos had been much overcut and the young shoots in consequence were short and thin. (p. 76)*

Beadon Bryant's Plan, 1896-1924

- *This [forest class] comprised forests stocked with sal and miscellaneous species of all ages, but having generally little sound mature timber left. The present division was placed in this class (p. 76)*

Champion's Plan, 1924-1938

- *This working circle contained all the sal and hill miscellaneous forests... and comprised 75% of the divisional area... Too much latitude was given to marking officers in deciding the exploitable diameters for sal, as the standard quality classes were not defined... Thinnings, were, however, not properly executed even in the area where the produce was readily saleable. (p. 79)*

- *The non-enforcement of lopping rules and... ruthless selection fellings, severe fires, and fellings for rights and concessions all combined to deplete the stocking of these forests. (p. 79)*

Coomb's Plan, 1939-1948

- *No marking rules were given, and these were left to be drafted by the territorial staff, resulting in indifferent silviculture work which, combined with overfellings for war purposes, seriously damaged the forests. (p. 80)*
- *No numerical limitation was placed on the removal of selection trees of sal, with the result that exploitation got the better of silviculture, and the existing percentage of sal already small was further reduced. (p. 81)*
- *Special war time fuel fellings were done in certain areas... reducing further the stocking of tree growth. Excessive grazing and lopping, wild elephant damage, and fire converted large areas into open grassland and valueless scrub of marorphalli. (p. 81)*
- *The adoption of a four year cycle [for cutting bamboo], though gave rest, did not solve the problem of overcutting because the rules were never enforced. The rule of leaving three year old culms was not enforced during the war period and even the new culms were cut... The sub-contract system of bamboo cutters... continued to do considerable damage. (p. 84)*

Gupta's Plan, 1949-1963

- *The subsidiary silvicultural operations, which were left to the discretion of the territorial staff, were either not done at all or were done when the crop did not get full benefit... (p. 87)*
- *Excepting for some canopy manipulation during the main fellings once in fifteen years, and that too only in mixed sal forest, no detailed prescriptions or suggestions to get sal natural regeneration were given. The natural regeneration of sal is thus generally deficient or totally absent except in some favourable localities... (p. 87)*

- *This working circle [sal experimental working circle, 2689 hectares] proved to be a complete failure as sal denova regeneration did not make any headway. (p. 88)*
- *Thus, many green trees were felled though they could more usefully be left behind for several years more. Exploitation had thus preference over silviculture, and the already small percentage of valuable species was further reduced. (p. 88)*
- *Nothing was done to check the growth of ravine formation and erosion of banks... the areas did afford grazing facilities, but at the end of the plan presented a deteriorated look owing to great damage by wild elephants in the already open and poorly stocked forests. (p. 89)*
- *The number of selection [sissoo] trees actually available was so small that in a few years the entire area was worked, and so commercial fellings ceased much before the expiry of the plan. (p. 91)*
- *The lopping rules were not enforced... As the allotment of areas changed from year to year, the Gujjars did not know the areas to which they would return next time and so felt no sense of responsibility and the damage continued unabated. (p. 92)*

S.S. Srivastav's Plan, 1964-1974

- *The progress of the sal coppice shoots has not been as satisfactory as envisaged in the plan... The regeneration of sal continued to be scarce except in some favourable localities... (p. 97)*
- *Enumeration was carried out only over six compartments of the working circle, which was of no help in forming a realistic idea of the growing stock present in the working circle. (p. 99)*
- *The regeneration of sal continued to be scarce, and that of other species excepting chir moderate to poor throughout the working circle. (p. 99)*
- *Though some teak plantations have suffered irreparable damage due to repeated fires... the results on the whole seem to be very encouraging. (p. 101)*
- *Prescribed fellings were done in the Gohri, Laldhang, and Kotdwara-kohtri felling series during the first year only. Thereafter, these were suspended in order to compensate for*

> the excess removal being carried out due to clear felling in the Bijnor Plantation Division.
>
> The four year [bamboo] cutting cycle appeared to be satisfactory. Cutting rules were, however, seldom observed as the contractors continued to work through sub-contractors, who in turn employed a large number of irresponsible cutters whose sole aim was to extract maximum possible quantities. As the result the crop has generally deteriorated everywhere... (p. 104)
>
> In actual practice prescriptions for this [lopping and grazing] working circle were never followed. More buffaloes than that prescribed were admitted on the plea that large numbers of Gujjars were displaced from the Bijnor Plantation Divisions and considerable area of Nalonwala block of Chani range had been clear felled for plantations. (p. 106)

The major destructive impact of forest management practices is evident from the above reports. Our field observations, as well as the comments from a wide range of Van Gujjars strongly reinforce this perspective, and directly implicate these practices as being the primary cause of forest deterioration. In particular, clear felling, over-exploitation and plantation development over one hundred years has reduced forest quality, removed nutrients, and threatened the ecological balance. The most destructive practices in this category were halted with the initiation of the proposed park in 1983. The results of these practices however, are fundamentally responsible for the state of the forests now. For example, there are plantations in Chilla which are entirely fenced off from grazing, yet the forest floor is nearly all covered by Latana weed which strangles other vegetation and inhibits forest regeneration. In another instance, there is a Van Gujjar dera near the border of the Bam (natural forest) and Rasulpur blocks (second generation plantation). Walking a short distance from the dera on the plantation side one finds a barren forest floor. In contrast, there is an abundance of natural ground cover if one walks the same distance into the natural forest, which has been lopped and utilised by buffalo for many years. There is a similar situation in the border area of the Luni Block, which was cleared and replanted with teak approximately twenty years ago. There is no undergrowth in the plantation and no evidence of elephants. A short distance away in mixed forest there is considerable undergrowth, extensive elephant dung, and evidence of elephant damage to trees. Both areas were open to grazing and lopping, although animals and people were not using the plantation strip because of the lack of fodder.

The Forest Department still permits the cutting of trees in specific circumstances, disregarding the protective policies. Thus, trees may be cut with permission for construction purposes within the proposed park, and we observed a number of such stumps. The Forest Department recently cut a large number of trees in Kanwli Park, (a unique forested area within Dehra Dun) in order to put in residences for employees, despite strong local protest (Joshi, 1995). The legacy of past practices remains evident.

(b). Illegal Felling and Tree Destruction: A second major factor contributing to the destruction of the forest is illegal felling and extraction. Van Gujjars and local villagers throughout the Rajaji area are united in identifying illegal cutting and destruction of trees as a regular systematic activity. In one khol, many Van Gujjars estimated that fifty truckloads of khair trees left the khol in a year. They indicated that a couple of trucks are caught by lower level forest officials in any one year, particularly when senior level officials are about to visit the area or are asking questions. The people caught return to their activities however, with impunity shortly after. Senior forest officials point out that penalties are very minimal and corruption is rampant in the judicial system, so as to make punishment ineffectual when people are caught.

Van Gujjars categorically report a wide range of incidents and illegal activities. They note in Kansrao that tractors are loaded with illegal wood by operators and passed through checkpoints. One of our research team, having lost his way in the core area of the proposed park, came upon people transporting illicit wood out of the forest. Van Gujjars note instances of illicit fellers socialising, eating, or drinking with forest officials in the proposed park. One frequently cited practice involves a person stripping a ring of bark from round a tree effectively killing it. The removal of the dead tree can then be more easily sanctioned. Van Gujjars in Ramgarh range report a standard practice in their khol in winter, whereby individuals strip bark off particular species of trees (padal and sain) which is sold for making alcohol at approximately 15 rupees a kilogram. They note that the individuals pay a fee for each tree, and that 20-25 trees are destroyed a year through this activity in their khol. In other instances, it is reported that labourers working in the forest are paid for their work in forest produce instead of their allotted wages. Senior officials note that criminal elements have prior knowledge of official visits to observe or investigate areas.

Van Gujjars state that they do not now report illicit activities because at best they are ignored and at most punished in one way or another. Specific examples included being charged with the offence they reported or threatened with fraudulent charges for other offences. One Van Gujjar related an example where an offence was reported, the accused person was released, and returned to ransack the dera. In another instance, a Van Gujjar man was shot in the shoulder after reporting a person for an offence.

By contrast, in the Chilla range, both Van Gujjars and local villagers indicated that illegal felling and poaching was less of a problem. These people noted that the Forest Department officials in the area often followed-up their complaints with enforcement efforts. This area was the exception rather than the rule.

The park plan does indicate that illegal felling is a problem and presents official statistics citing 6272 cases in the last seven years. Berkmuller (1986) points out that these statistics are likely to be underestimations if some officials reporting cases were involved in them. Park authorities indicate that government servants have been charged in some cases. Although illegal felling is admitted to be a problem in the park plan, there is little discussion of how it should be addressed in the strategies section, beyond suggesting there should be more guards to improve enforcement. Van Gujjars and villagers argue that more guards would not decrease the illegal activity.

Local peoples report that there are both individual thieves and organised gangs. Lower forest officials are poorly paid, live for long periods away from their homes and families, with little remuneration, and nothing more than an official duty to protect the forest. Their self-interest, under very difficult working conditions, can become making money through accepting payments. In addition, in some instances they are threatened by the criminal elements, so that they have no choice but to condone or support the illicit activity. Attempts by local people to report offences and offenders may result in punishment of the informant, so silence is enforced.

These illegal activities represent a major problem for a system which is overworked with few resources and spread over a large area. It is not a problem for which individuals can or should be blamed, it is a question of structural problems of a complex nature which bring pressure on different individuals in different ways, giving individual officials no effective means to fight the problems. Local people do not assist them.

(c). Van Gujjar Impact: Park authorities have pointed to Van Gujjars as the foremost cause of forest destruction: "Van Gujjars enjoy the distinction of being the most destructive to the forests" (Kumar, 1995, vol. 1, p. 217). It is important to evaluate this charge. Van Gujjar's' use of trees includes lopping for fodder and use of dead wood for firewood and dera construction. Their impact in each area will be examined in sequence.

Lopping: Van Gujjars are reported to ignore all lopping regulations, and lopping is viewed as destructive to the forest, i.e., "heavy and concentrated lopping is resulting in trees drying and dying (ibid., p. 178). No evidence nor concrete examples are cited to support these claims, and we did not observe large numbers of dead or dying trees. If the claims are accurate and trees have been lopped for generations, one would not expect any trees to have survived. Contradicting this claim, our surveys suggest that forest blocks in the central areas of the proposed park where the highest numbers of Van Gujjars reside are of higher quality than the border areas (see Map 5). One Van Gujjar noted: "If we do not stay in the forest, then there will be no forest."

It is important to examine the available research on the effects of lopping in order to evaluate the impact. No overall conclusions are evident on the impact of lopping as some studies demonstrate an adverse impact (Rawat, 1993) and others do not (Deb Roy et. al., 1980). The most recent study in the Rajaji area (Edgaonkar, 1995) compared two adjacent and ecologically similar sections of forest, one with moderate lopping and one with no lopping, in the Dholkhand range. This study found that the lopped area had more growing seedlings and fewer weeds, but the authors were at a loss to explain results contrary to the initial hypothesis. We asked a group of Van Gujjars for their interpretation of the results, and they suggested that buffaloes' presence in the lopped area returns nutrients to the soil through manure (a hypothesis supported by Cincotta & Pangare, 1993) and disperses seeds. Moreover, they explained that Van Gujjars clear weeds beneath trees when they lop.

Clark et. al. (1986) conducted the only other lopping study in the Rajaji area and found that Van Gujjars lop trees in a sequence parallel to leaf fall. Thus they lop each species just prior to its leaf fall in the winter season. Within species, we observed that they lop trees first at higher altitudes which have less water and earlier leaf fall than to trees in lower and wetter locales. Sharma & Gupta (1981) found that the time of year of lopping is critical in determining the amount of impact on the tree. Trees lopped in the winter season before leaf fall grew more and produced more fodder than trees lopped in other seasons. Indigenous practice and scientific knowledge support each other.

There is a clear difference between Van Gujjars and park officials about appropriate and effective lopping. Official regulations indicate that the top third of the canopy must be left and no branch cut which is larger than 2.5 centimetres in diameter. Many Van Gujjars indicated that the top main branches must be left with some leaves throughout, though not necessarily one-third of the canopy as in the regulations. They disagreed with the notion that all branches above a small diameter must be left, stating that lopping is akin to pruning and that one shapes the tree according to light and growth patterns. Thus, one should thin out the branches to encourage growth while leaving select branches in strategic locations. Ghulam-Ud-Din from the Mohand Range noted, "We believe in careful lopping, we don't fell trees but bring them up so that we can have more fodder." Van Gujjars indicated that it was preferable to lop only every other year unless there were excellent soil conditions, but some through that there were not enough fodder trees available to allow for the rest year. They judge carrying capacity of an area in relation to the number of fodder trees available so as to only lop every other year. They pointed out that lopping opens up the tree canopy, and allows more light to penetrate to the forest floor, which in turn promotes more undergrowth (supported by Clark et. al., 1986) and seed regeneration.

A team consisting of A.Clark, H.Sewill and R. Watts in 1986, found "that crown cover was relatively unaffected by lopping and that there was an increase in ground vegetation in an area with lopped trees which would decrease the possibility of erosion... The study team saw no evidence of dead or dying trees other than those knocked over by elephants. The practice of lopping is very similar to that of shredding carried out on trees in some areas of England in the Middle Ages until it was found that coppicing the tree (that is, cutting off at ground level) produced the same amount of regeneration with a lot less effort... The theory of the 'Tragedy of the Commons' is not applicable to the Gujjars' situation: land is apportioned to each family and boundaries respected. Families practising transhumance always return to the same area year after year.

Our field observations throughout the proposed park found that many trees were lopped in accordance with Van Gujjar ideas on proper lopping, although this did not conform with official regulations. We also observed some lopped trees in which only the topmost extension of main branches had leaves left, contrary to stated Van Gujjar perception. This practice was acknowledged by Van Gujjars themselves as occurring in some cases. The more intensive lopping appeared to be associated with the amount of insecurity felt by Van Gujjars about their tenure on the land. Lopping appeared to be complete among the more insecure Van Gujjars. By contrast, one Van Gujjar in the Mohand Range stated "If forest management is given to me, I will grow fodder trees and I won't need external aid like wells and saplings. I can invest in it like it is our own land."

Van Gujjars also noted that when faced with fodder scarcity, some Van Gujjars, particularly those with more financial resources, pay fees to forest officials who allow them to lop illegally in closed khols or on prohibited species. Edgaonkar (1995) noted lopping of sal trees in the summer months among Van Gujjars who remain in the proposed park during this season. It is not an inappropriate time to lop sal as this species loses its leaves twice a year, regrowing them in the monsoon. It is however a prohibited species for lopping because of its commercial value. Interviewees also noted that the forest officials would allow some settled Van Gujjars, in return for a fee, to move their buffaloes in the summer months to other khols which had been vacated by nomads, effectively robbing those khols of the needed regeneration period. This system results in some Van Gujjars not having to live within the limits of their areas, while other Van Gujjars feel powerless in the face of the Forest Department authority which supports the activity. Whereas the Van Gujjar Panchayat system was reported to work effectively in disputes outside Forest Department related issues (e.g., domestic disputes), some Van Gujjars noted that the Forest Department could overrule the panchayat in forest issues, through granting favours and protection to a select few.

As with other issues, such as illegal felling and villager use of the forest, there is no effective enforcement of regulations regardless of their merit. Many Van Gujjars report that the only result of any restriction on forest use is to increase the market price of the fee. Unfortunately, the final cost is borne by the forest, to the extent that paying more unofficial fees puts more pressure on Van Gujjars to generate income and keep more animals.

In short, there is no evidence to suggest that lopping is an inherently destructive practice generally speaking within the proposed park area, although there are problems in particular areas where there is poor forest. There are differences in forest in across the proposed park but they are not correlated with the amount of lopping nor the number of Van Gujjars. Van Gujjars noted that the impact of lopping on a tree was a complex topic which depended on soil type, water scarcity, species type, age, exposure, and other ecological factors. For example, two trees of the same species (Bahera) and similar ages growing within fifty meters of each other in the Luni Block of the Chilla Range are photographed. The tree in the first photo had been lopped in the previous year while the tree in the second photo had not been lopped for years because it was on the boundary line between a Van Gujjar and villager zones of control. It is quite clear that the lopped tree is more vigorous in leaf growth and in a better shape.

Van Gujjar ideas on lopping management are based on extensive indigenous knowledge, which is supported by available scientific knowledge. Van Gujjars identify that the present forest management structure is ineffective, and that negative impact of Van Gujjar lopping, where they exists, is a product of individuals attempting to survive within a situation where they have no power to make systematic changes. They are quick to suggest solutions for dealing with existing abuses, if they were to have the responsibility of policing the forest, and the practices of their peers.

Van Gujjar Use of Firewood : Van Gujjars utilise wood from the forest for fires for cooking and heating. The park authorities state that, "a Van Gujjar family recklessly consumes on an average about one quintal (100 KGs) per day of fallen fuel in the lopping areas on the pretence of protecting their cattle from carnivores and their self-protection from cold" (Kumar, 1995, p. 213). On that same page however, a figure of

20 kg per day is noted, and elsewhere the report cites figures of 40 KGs a day (ibid., p. 173) and 10 KGs a day (ibid., p. 235). Whatever the accurate figure, the wood is not used as a "pretence" for keeping warm. It *is* used to keep warm, which is a legitimate function of a fire in the winter.

Our field observations indicate that 100 KGs of wood per day is grossly exaggerated. Clark et. al. (1986) reports up to 40 KGs a day (which suggests a lower average), and this would be in accordance with our informal observations, possibly 25-30 KGs per day. Our field observations indicate there is no shortage of dead wood around, and Van Gujjars report no problem in locating dead wood close by, unlike Himalayan villages where women walk long distances for firewood due to deforestation. We frequently observed fallen dead wood in significant quantities lying within one hundred meters of deras. Elephant damage to trees is one prominent source of dead wood. There is not a shortage of dead wood nor an evident trend towards a depleting supply. Of course, wood that is burned is not returned to the soil through decomposition, which is a loss to the ecosystem. Van Gujjars are a part of the ecosystem however and it would seem that they have a claim to using its resources in a sustainable way for basic warmth and survival.

Van Gujjar Dera Construction : A second Van Gujjar use for wood comes in the construction of deras. If a Van Gujjar family is able to return to the same place for many years, then the wood for a dera would have to be replaced only every five years, and the major beams every ten years. In fact, if Van Gujjars leave the dera empty through the summer season, the materials are either sold by forest officials or removed by villagers so that the dera must be rebuilt the next season, requiring more wood. As a result, Van Gujjars have increasingly adopted the practice of leaving at least one family member behind to protect the dera from destruction. Thus, a Forest Department management practice results either in the needless reconstruction of a dera, or in the need to leave people behind in the summer, increasing the pressure on the forest.

Van Gujjars explain that they use only dead and fallen wood for dera poles, unless in emergency a branch might be cut off a tree. The thatch must be replaced each year. Trees are never felled. This was substantiated by our visits to a number of recently completed deras, as well as several others in the construction phase. The poles in these structures were clearly seasoned dead wood. In short, a Van Gujjar family uses available dead wood once every five years, assuming they are not moved, to construct a modest hut. This impact should be assessed in relation to the magnitude of other impacts and the rights of indigenous people to a home in the forest.

(d). Industrial and Urban Encroachments: These encroachments have taken the form of the various industrial and government land uses that have replaced forest land for several decades in the past. The major impact is reduction of available forest habitat, both for wildlife and the use of local people. The new Haridwar District Headquarters is a particularly glaring example, in that it involved clearing a large tract of land, while the government was at the same time attempting to evict local people from forest. In general the loss of forest land around the edges of the proposed park concentrates both the wildlife and the Van Gujjars in smaller areas and increases the biotic pressure in those areas. This has been a long term trend with the development of land in the proposed park region over the last few decades. The park authorities have either been unwilling or unable to challenge industrial and government interests.

(e). Villager Impact: The main use of wood by the local people living on or near the edges of the proposed park is for fuel. With a large population around the proposed park and a high level of dependency, it seems clear that the amount of fuel wood removed from the forest is immense yet very difficult to estimate. In many instances fuelwood was a traditional right until it was restricted by park authorities. Berkmuller (1986) counted headloads of wood moving out of the proposed park at specific observation points and noted between 50 and 100 headloads per hour (roughly 35 KGs each)

in a majority of fifteen observation points. Park statistics estimate that 20,000 men and women collect firewood from the proposed park on a daily basis, although it is unclear as to how this figure was calculated. Berkmuller (1986) mapped the areas from which firewood was collected in the western half of the proposed park based on discussions with forest guards. In most areas, villagers were penetrating from 0 to 5 kilometres into the proposed park, entering only as far as five kilometres in some specific locales. Map 7 indicates the areas of Van Gujjar and villager use based on extensive discussions throughout the area with Van Gujjars and villagers during the development of this plan.

The amount of extraction is clearly much greater than that taken by the number of Van Gujjar families collecting wood in deeper regions of the proposed park, although the latter group received more emphasis in park reports. It is also unclear how much of the wood collected is for local consumption and how much is sold commercially in cities and towns. Both Van Gujjars and Berkmuller (1986) report particularly high levels of extraction in khols adjacent to Haridwar where the fuelwood is likely being used or sold in the city. As with other illegal items extracted from forests, a person may pay a fee to the forest officials for collecting firewood, even though it was a traditional right. Van Gujjars reported that a headload would cost 10 to 15 rupees. Without any clear information or effective set of regulations, it is impossible to ascertain the level of forest impact. The forest area becomes a common land resource for large numbers of people and, since each person pays for the privilege, no one feels responsible for the resource. This lack of community responsibility for discrete areas has been identified as a primary cause for village forest degradation for several decades.

Summary of Contributing Factors in Forest Deterioration: It is evident that forest management practices are the major long term factor contributing to forest deterioration, whilst illegal felling is an the most prominent factor at present. Lopping is not necessarily a destructive factor, although it may be having a negative impact in some areas for reasons related to both

Van Gujjar practices and Forest Department shifting of Van Gujjars. Van Gujjar use of wood for fuel and dera construction has some impact but appears minimal in relation to other factors. Industrial and urban encroachments make an impact on the forest, in that they reduce the amount of forest land available, and restrict the habitat for both wildlife and forest dwellers, increasing the biological pressure in the remaining areas. There appears to be a large but unknown level of pressure on border areas, due to the removal of fuelwood by villagers.

Depletion in Ground Vegetation

Status

There is no detailed comparative data which establishes if and when there has been depletion in the ground vegetation over the years. Based on extensive time spent in the proposed park, we have observed that there are significant discrepancies in ground vegetation across the area. Some areas are quite barren while others have good ground cover. Park authorities suggest that overgrazing is depleting ground cover, but there is no indication of specific problems by area. The review of past working plans indicates that grazing has been going on for generations, although authorities blame the general situation on relatively recent increases in Van Gujjar and villager animals. It is very difficult to ascertain however, what has been an on-going part of the Rajaji ecosystem in terms of the presence of cattle, and what may be increasing and recent deterioration in ground cover.

Analysis of Contributing Factors

(a) Forest Management Practices: The history of destructive forest management practices with respect to timber exploitation has already been described in detail, and it clearly results in the loss and degradation of ground vegetation as well as diminishing the quality of the forest. Plantations are particularly destructive of eliminating ground vegetation. Taungya plantations in particular remove additional nutrients in the initial years through crop production.

103

Another major area in which forest management practices have affected ground vegetation involves the extraction of bhabbar grass. Bhabbar grass is utilised by poor villagers along the edges of the proposed park, predominantly on the southern fringe, to make baan rope which is their primary source of income. The grass grows on the ridges of the Shiwaliks in open tracts. Villagers and Van Gujjars report that the total crop of bhabbar is being cut each year, and that continuous exploitation without rest is a drain on the ecosystem and the ground cover in time. On the other hand, villagers, Van Gujjars, and research studies indicate that well-managed and sustainable extraction of Bhabbar is not an ecological problem and, in fact, removing the dried grass is a means of reducing the potential of destructive forest fires (Poffenberger et. al., 1995).

The park report cites a Baan Industry report from 1985 as indicating that 10,000 families work on a full-time basis in baan production, while VIKALP (1994) found that there were approximately 3500 families dependent on Bhabbar in twenty-one villages on the southern proposed park border. The percentage area of the VIKALP study of the whole Baan area is such that the two estimates seem roughly compatible. VIKALP (1994) estimates that each family requires sixteen quintals of grass if one assumes an eight month season.

Villagers identify the history of forest management practices as the major contributory factor to the deterioration of the resource through the years. They report that access to bhabbar grass has been a traditional right for generations that was continually recognised under British rule. Villagers personally cut the grass in the park area. The Indian government after independence however, altered this system in the early 1950s as it took increasing control of the forests. The Forest Department moved to a system of selling the resource to private contractors, who in turn were responsible for removing the grass and selling it to villagers and the paper industry, a second major use for the grass. This system evolved into the major contractor is hiring a smaller contractor to remove the grass from a given set of khols. The smaller contractors increasingly hired outside labourers to do the cutting in preference to local people. The smaller contractor sold the grass to the major contractor, who in turn sold it to another smaller contractor, who was responsible for selling it to the local villagers. The Forest Department and the contractors each took a share along the way resulting in the villagers paying extremely high prices. The availability of the paper industry to take as much grass as possible was an incentive for over exploitation and financial gain by all, and contractors cut even those young tender shoots which are particularly important for regeneration.

Political pressure and a human outcry with respect to the exploitive nature of this system resulted in the government agreeing to set up a state forest corporation in 1982 which was to remove the process from the private contractors. It was also agreed that at least 50% of the workers would be from the local villages. In fact, the Forest Corporation (Van Nigam) simply took over the role of the major contractors, while the small cutting contractors and the distribution system remained in place. Van Nigam paid the Forest Department for a targeted amount of grass to be cut. The local people were still not hired as the actual cutting was managed by the same small contractors as before.

Van Nigam was responsible for bhabbar extraction in the proposed park area up to 1986 and remains responsible in the Mohand area and Shiwalik Division up to the present. The cost of bhabbar for villagers is now set at 180 rupees a quintal (100 KGs) while smaller pieces of grass are sold to the paper companies at 60 rupees a quintal. The price for villagers remains high as Van Nigam charges include overheads, the Forest Department requires a fee for the access, and the contractors are paid for their efforts. At present Van Nigam is trying to raise the price to 280 rupees a quintal despite villager protests. If a villager were to cut the grass personally, a quintal would be valued at approximately 90 rupees for the labour (45 rupees a day). In addition, Van Nigam only auctions the cutting rights for a three month season whereas grass is traditionally cut over six to eight months. As the contractors however, know the areas in which they will be cutting for Van Nigam when the season opens, they move into these areas three months early and cut grass illegally in return for paying fees. This grass is then sold on the open market at 350 to 400 rupees a quintal. In short, it is to the advantage of the contractors to cut as much as possible before the official season opens.

Whereas the above system still operates outside the official proposed park boundary, late in the 1980s legal extraction of grass from the park area was reduced, and then banned. This resulted in a series of confrontations with local villagers in the early 1990s in which promises were made for villagers to have access to alternative resources, that never materialised. Instead, the situation has evolved into continued complete exploitation of the resource by villagers in return for a payment of 20 rupees a quintal (roughly two headloads). From a villager perspective, other than being illicit and subject to Forest Department authority, this system is an improvement over the forest corporation system as they are able to extract the grass themselves at far less cost. This attractive system has now resulted in one of the more powerful grower contractors hiring villagers to cut additional grass through the headload system, and then sell it to more distant villages or commercial interests. The net result is over exploitation of the resource. Van Gujjars report from 100 to 200 headloads of grass leaving individual khols in a day in the Dholkhand range. Van Gujjars also report that grass is being transported to northern border villages across the Ramgarh range, though in lessor quantities than to the south. It is not the major economic activity of these communities, and it is unclear how much of the grass is made into baan locally and how much is sold to commercial interests.

In reviewing this complex history, the pattern of over-exploitation which has been destructive to the ecosystem is evident. At a simplistic level the Forest Department can cite local villagers for extracting the grass, but a more detailed and sophisticated analysis indicates that it has been the management practices and incentives that have resulted in over-exploitation while also exploiting the local people dependent on the resource. One Van Gujjar reported challenging a forest guard when he refused to deal with the high level of bhabbar extraction in his khol. The guard responded by saying that he required the fees for the bhabbar as he in turn had to pay Rs 1000 to his seniors in order to maintain his position in the block. These patterns of exploitation are again a systemic problem for which individuals cannot be blamed.

(b). Van Gujjar Impact: Van Gujjars are blamed by park officials and conservationists for increasing their animal and human population, which puts particular grazing and biotic pressure on the park area. It is also claimed that large numbers of Van Gujjars are not nomads now but present for all twelve months in the park area, increasing pressure further. It is important to examine these issues carefully to determine the level of pressure and the reasons for it.

There is a consensus in the research literature that over-grazing is destructive to the ground vegetation, reduces biodiversity, alters the plant species make-up, increases the presence of non palatable weeds, suppresses plant regeneration, and increases soil erosion by reducing ground cover (Dwivedi, 1992). Goats, sheep and camels are significantly more destructive than cattle due to their ability to consume a much wider diversity of plants. Properly managed grazing however, is reported to be beneficial in that moderate browsing promotes plant growth, cattle disperse seeds, and contribute nitrogen to the soil in an easily accessible form through manure (Goudie). Grazing or grass cutting has been re-introduced in some protected areas due to their beneficial effects (Kothari et. al., 1995).

The impact of grazing must also be considered in relation to climatic changes through the seasons. The monsoon initiates an explosion of plant growth so that the area can support a high level of grazing, while still leaving ample vegetation for other creatures and forest regeneration. During the winter season plant growth slows but continues, with the pinch period coming in the summer season when growth stagnates. Thus an area can sustainably support a higher number of cattle in the monsoon and winter seasons than in the summer season. With respect to Van Gujjar impact, the

107

issue of concern is the summer season as buffalo prefer, and are almost totally dependent on, tree fodder obtained through lopping in the winter. After leaf fall, there is no alternative for grazing within the ecosystem if the buffaloes remain in the area. Yet if buffaloes are nomadic in the summer and monsoon, the grass crop regenerates and is well stocked at the onset of winter, increasing the carrying capacity. In short, grazing is not an inevitably destructive practice and has been a part of the ecosystem for generations. It must be effectively managed however, in relation to carrying capacity, with particular attention to seasonal growth patterns and fodder availability. Analysis of Van Gujjar grazing must be separated into winter and summer impact, while remembering that summer impact prevents regeneration and affects capacity the next season.

Our observations in the proposed park area in the winter suggest that there are khols where overgrazing is evident, and others where it is within reasonable limits. Extensive interviews with Van Gujjars in nearly all khols reinforce this view. Van Gujjars are clear in recognising that there are too many buffalo in some khols for the present forest quality. In other areas they report a reasonable balance or additional forest resources. Park authorities blame an increasing population of Van Gujjars, yet they have not even doubled assuming 512 families in 1937 with 6 persons a family and approximately 5500 Van Gujjars today in the recent census. Although their birth rate has been high, so has infant mortality. In contrast, census figures indicate that the population of the Dehra Dun district multiplied by roughly four times between 1941 and 1991, from approximately 2.5 to 10 lakh people. Simultaneously, there has been a significant deterioration in the forest due to forest management practices, and the Van Gujjars have been powerless to make decisions to improve these practices to compensate for population increases.

With respect to winter management, Van Gujjars are able to hold more buffaloes within their area than permitted in return for paying a fee. This is a legitimate example of negative Van Gujjar impact but the responsibility for the impact is jointly shared with the forest officials who sanction the process. The system is unable to enforce regulations. Van Gujjars see the forest deteriorating around them due to other management practices beyond their control, and therefore decide to use the forest themselves rather than watch others use it. There is anger from a broad spectrum of Van Gujjars toward the favouritism by some forest officials toward a select few of the more affluent Van Gujjars. These select Van Gujjars advocate Forest Department policies and carry out favours in return. For example, select Van Gujjars are paying fees to reside or lop in "closed" khols while the Van Gujjars originally residing in those khols have been moved into already fully occupied khols, raising the number of buffaloes above the carrying capacity in these khols. Again there is no incentive to live within sustainable forest limits. But despite these factors, there are only problems in the winter season in some khols. The study demonstrating more seed regeneration and less weeds in a lopped versus a non-lopped area attests to the fact that Van Gujjar occupation of areas can be a benefit to the forest (Edgaonkar, 1995).

Summer negative impacts are more substantial and widespread. Ranipur, Motichur, and some areas of the Kansrao are areas where Van Gujjars report particular stress (see Map 5). These ranges have higher proportions of non-nomadic families. Forest Department management practices have been a primary factor for the demotion of nomadism for a number of reasons :-

1. *Hill Encroachment By Landowners:* The most frequently cited Van Gujjar disincentive for going to the hills is the encroachment by powerful landowners into Van Gujjar pastures there, and the increasing difficulties in finding suitable stopping places with fodder in reserve forests on the migration route due to similar encroachments. In both of these instances the encroachment is illegal but may be sanctioned in return for fees to some forest officials. Moreover, Van Gujjars report that forest officials in local areas side with local people in disputes, so that the Van Gujjars have no recourse to an effective justice system. In recent years, wealthy landowners with political influence have also begun contracting with local

people to maintain large herds in the highland pastures. Recently however, there has been some change in Forest Department policy, and DFOs in the region were requested to take Van Gujjar complaints seriously. In several instances this has paid off, with Van Gujjars reporting support from DFOs when they were faced with harassment.

2. *Forest Insecurity:* Van Gujjars have felt particular pressure to remain in the Rajaji area in recent years, because of the increased shifting of permits by forest officials in return for fees. Some Van Gujjars report that they have lost part of their forest permit while they are in the hills, and hence do not want to leave their forest unoccupied. There is also the fear based on prior threats by forest officials that they will not be allowed back in the park area if they depart in the summer. This fear became a reality in 1992 when they were initially forbidden re-entry. This order was overturned only by massive protests, publicity, and the threat of roadblocks. In the interim period some buffaloes died while awaiting re-entry.

3. *Land Offers:* In the 1970s the Forest Department proposed that it would give a large number of Van Gujjar families in the Ranipur, Motichur, and Kansrao ranges agricultural land in return for surrendering their forest permits in the Uttar Pradesh Hills in the summer. They were to still have access to the Rajaji forest area in the winter. 157 families on a Forest Department list were promised land but, in fact, only approximately one-third of these families received it, and no one was given ownership papers. Some of the land was within the present proposed park boundary, and some outside it. With the initiation of the proposed park in 1983, however, those within the boundary lost their right to the land, and numbers of those outside were given a choice to have either the land or the forest inside the proposed park area. Most of them returned to permanent occupation of the forest, as they could not sustain themselves on land alone. The net result of this process was to settle a large number of Van Gujjars in the park area while they lost their permits to go to the hills.

4. *Summer Forest Transfers:* Non-nomadic Van Gujjars from khols which cannot maintain their herds in the summer (including those settled by the 1970s scheme cited above) are moved into the khols of nomads who have gone to the hills in the summer (e.g., Bam khol). As the result, the nomadic khols do not regenerate, and the benefits of a sustainable forest are lost to the nomads. It makes more sense to remain behind if the forest will be used anyway, and be able to respond to the political context. The net result is more Van Gujjars in Rajaji as the non-nomadic Van Gujjars remain regardless.

5. *Dera Destruction:* Another factor influencing Van Gujjars to remain in the Rajaji area is the reality that their deras may be auctioned off by forest officials or pilfered by local people if they leave them unoccupied.

In brief, some aspects of Van Gujjar behaviour have had a negative impact on ground cover in some areas, particularly in the summer. Unfortunately forest management practices have provided a disincentive to nomadism which increases ecological pressure in Rajaji. Despite these issues, the fact that high numbers of Van Gujjars have been living in the best Rajaji forests over many years, and yet they are still the best forests, indicates that their presence overall has not been detrimental. Van Gujjars in fact, argue that their presence protects areas from far more destructive influences, and results in a net benefit to the forest. The primary issue is the need to reduce summer impact, in order to allow the forest to regenerate in keeping with traditional practices.

(c). Villager Impact: Grazing has been a traditional right of the peripheral villages for generations and takes place in a 0 to 5 kilometre band around the border of the proposed park (see Map 7). Census figures indicate there are 77,000 villager animals in the peripheral area and park records permit up to 64,000 animals for grazing. The proposed park plan implies that the permitted level is the carrying capacity (p. 177) but does not explain if or how the carrying capacity was calculated. [Van Gujjars typically define carrying capacity based on a combination of tree fodder and grass resources, assuming that trees are lopped every other year and that

buffalo eat approximately 9 KGs of grass a day when grazing]. Despite park officials' emphasis on the damage caused by Van Gujjar animals, villager animals clearly put dramatically more pressure on the areas that are utilised. There are up to 4 or 5 times as many villager cattle using approximately 25% of the area (generally the periphery) as the number of Van Gujjar buffalo using roughly 75% of the area (generally the core area, see Map 7). In addition, villager livestock do not utilise lopped leaves in significant quantities, depending almost entirely on grass, and are present in the area twelve months a year. Moreover, since plantations are more dense in these border areas, there is less ground vegetation available. On the other hand, all the cattle in the area are not using the proposed park. Our field observations clearly indicated that there is significantly less vegetation in the border grazing zones, and what is present is severely grazed. It should be noted that there is no system in place to encourage villagers to enforce limits to grazing or to define appropriate zones. Villagers recognise the limitations on the resource and yet, if a given village shows restraint, the excess will simply be utilised by livestock from another village.

Another significant grazing impact is the use of khols in the Chilla area by large herds of sheep from the Uttarkashi Hills in the winter season. Herds of 500 to 1000 sheep come into the area and spend a week or two in a given khol before moving to another one. We observed one such herd estimated to be between 600 and 700 head. This had been a traditional practice which was banned in recent years by park authorities. It continues however, in return for fees given to park officials. Another smaller grazing issue is the presence of camels in the Mohand range which are used for the transportation of bhabbar grass by contractors. Camels eat far more than cattle or buffalo, and from a far wider range of vegetation. The amount of pressure resulting from this practice is unclear. Bicycles or headloading are alternative means to camels transporting grass out of the khols.

Extraction of gorla grass is another resource removed from the forest by peripheral villagers. Van Gujjars report that this practice is recent with the advent of the fee system for bhabbar extraction. Gorla is now removed in addition to bhabbar while under the previous contractor system the focus was on bhabbar extraction. It is unclear how much is removed or whether there is a significant impact.

To sum up, there does appear to be overgrazing in the boundary areas seriously depleting ground vegetation, largely because there is no effective management system for grazing. There is no means which villagers can take responsibility for the resource which has been their traditional right.

(d). Industrial and Urban Encroachments: The impact of this encroachment is again straightforward. Less land available due to development within and outside the park area reduces the overall level of resources ,and increases pressure within the remaining areas.

Synthesis of Factors in Depletion of Ground Vegetation: The analysis of contributing factors to the depletion of ground vegetation again demonstrates major problems in the management system whether one is referring to bhabbar grass extraction or grazing. There is no evidence to suggest that moderate resource extraction is harmful, there is however evidence of over-exploitation. In particular, the existing management system has provided incentives for over-exploitation of the resources over decades while preventing local people from taking any responsibility or being accountable for the resource. Even under the present system, though Van Gujjar animals show some negative impact, it appears that Van Gujjar presence can be a net benefit to the areas in which they reside.

Weed Infestation

Status

Table 8 presents the data on weed infestation for the ranges in the park area. Clearly weeds are a major problem throughout all ranges, with infestation ranging from 28% of the area in Gohri to 48% of the area in Motichur. The average range has 38% of its area infested.

Analysis of Contributing Factors

(a). Van Gujjar Impact: Park reports attribute weed infestation to overgrazing by Van Gujjars and villagers, but the discrepancies in percentages across ranges and khols do not support this contention. Gohri has the lowest percentage infestation of weeds yet has a very high villager population, and is proposed for removal from the core area as the result in the park plan. Dholkhand block, the one which has been closed to grazing for many years has a similar percentage of weeds as the rest of the proposed park. A study by Edgaonkar (1995) in fact, found a lower number of weeds in a moderately lopped Van Gujjar area than in a closed unlopped area with comparable terrain.

Bhatnagar (1991) found that Van Gujjar lopped areas had more shrub undergrowth, but that this undergrowth was not infested by weeds. Van Gujjars report that they weed as part of their lopping activities, although they indicate that their weeding depends on the amount of security a person feels with respect to land tenure. One Van Gujjar in the Mohand range has weeded his entire permit area, and has encouraged his neighbour to carry out the same practice so that both areas remaining weed-free with minimum effort. On the other hand, Van Gujjars report that they usually do not weed in areas where they know they will not be returning.

(b) Forest Management Practices and Villager Impact: Forest management practices over the years have resulted in a deterioration in the forest and an increase in plantations. Weeds proliferate under poor forest and soil conditions compared with other plant growth. Forest management practices have combined with over-grazing, particularly in the border areas, to bring about an even greater increase in weeds. Thus, clear felling and over exploitation opens up the canopy to encourage undergrowth while over-grazing eliminates plant diversity, allowing weeds which are not eaten by livestock to flourish (Prasad & Bhatnagar, 1988). Once weeds arrive they spread quickly and easily unless action is taken against them. The Forest Department has not had the resources to take on this task, and villagers do not have responsibility for the forest areas they use, and therefore see little reason to weed. There have been no Forest Department efforts to work with local people to reduce weeds. Moderate grazing in a healthy forest need not result in weed infestation.

Table 8
Weed Infestation in the Proposed Rajaji National Park

Range	Total Hectares	Weed Infested Area in Hectares	Percentage Infected
Kansrao	7933	3695	47%
Motichur	8042	3859	48%
Ramgarh	7703	2831	37%
Chilla	14829	4896	33%
Gohri	10177	2816	28%
Dholkhand	13229	4821	36%
Chillawali	11531	4239	37%
Ranipur	8525	2625	31%
Total	**82042**	**29784**	**37%**
Dholkhand Block	1479	563	38%

Excerpted from Kumar (1995)- *Management Plan of Rajaji National Park, Vol. II*, p. 42-43.

Synthesis of Factors in Weed Infestation: Forest management practices and over-grazing in border areas have been key factors in weed proliferation. Van Gujjars presence appears to be associated with lower levels of weed infestation. Van Gujjars, as well as villagers, have the potential to play a much greater role in the reduction of weeds if they feel secure in their land tenure.

Wildlife Problems

Status

The clearest means to define whether and how much of a problem there is with wildlife is to examine changes in populations over a period. Official wildlife statistics are provided in Table 9, although the time between surveys is short, making trends more difficult to verify. There were decreases in animals with the exception of elephants, panthers, and neelgai, unless the methods are unreliable and the results are due to sampling error. In particular there are decreases in most herbivores, but not elephants, which is in keeping with extensive reports of systematic poaching. If these decreases are reasonably accurate, the huge loss in species such as sambar could not be explained by gradual habitat deterioration through the years. Poaching is the only plausible explanation other than method error in the census. There seems little explanation except method error for the dramatic increase in panthers and it is unclear as to which figures are more reliable. The 1993 figures provide a breakdown by sex, suggesting more accuracy.

Table 9
Wildlife Census Figures For the Proposed Rajaji National Park

Animal	1989	1992	1994
Elephant	338	346	552
Tiger	40	---	17
Panther	66	---	110
Sambar	---	1007	801
Chital	---	4619	4593
Kakar	---	585	271
Goral	---	270	148
Para	---	13	2
Neelgai	---	310	416
Pig	---	1419	988
Monkey	---	4404	3902
Langur	---	3847	3207
Bear	---	9	7

Data from published park reports

The park director particularly noted that the 1993 elephant statistics were more reliable than the earlier figures, because they were based on visual sightings (Ahuja, 1994). Unless there was a dramatic influx of elephants into the area in one year, the 1993 increase must largely be due to the difference in census methods. A dramatic one year influx seems unlikely given the well documented obstacles to elephant migration in the region. Census figures for the Chilla (Lansdowne) area however, were 192 in 1976 (Singh, 1989b) whereas they are now estimated as approximately 350 according to the park plan. This suggests that there has been an increase with time. Villager reports of increased crop damage due to elephants may also indicate an increase, although this could also occur as the result of habitat restriction.

Analysis of Contributing Factors

Diminishing numbers of wildlife could be explained by predation (natural or illegal poaching), food and/or water scarcity, or loss of habitat. Each of these issues must be examined in relation to the contributing factors previously identified.

(a) Illegal Poaching: Both the statistical data on wildlife trends above, and the results of interviews with a wide range of Van Gujjars and villagers, suggest that this is the primary issue in restricting wildlife populations. Elephants in particular, have not been poached to date, and have increased in numbers over the last few decades. By contrast, the numbers of other herbivores, for which poaching is widely reported, have apparently diminished. If the major factor was reduced food or water supplies, then one might expect that the numbers of all sorts of herbivores would exhibit similar trends. In addition, tigers, a particularly valuable species for poaching, have decreased while panthers have not.

Poaching was reported as a major problem by nearly every Van Gujjar that we interviewed. The Ramgarh and Kansrao ranges have the highest levels of poaching according to Van Gujjars, but there is also obvious and regular poaching in the southern ranges of Chillawali, Dholkhand, and Ranipur; Chilla Van Gujjars reported less poaching. We observed relatively high numbers of chital during our visits to the Chilla range. In Kansrao, Motichur and Ramgarh, Van Gujjars noted that poachers work on a regular basis, using packs of hunting dogs, and paying a regular fee. They stated that these areas had much fewer animals, both because of the poaching and because animals remained in the southern areas where there were more Van Gujjars and less poaching. They reported that herbivores moved close to Van Gujjar deras in the night throughout the area it was safer for them. We observed deer around deras in the evening several times during our field visits. Van Gujjars reported that there were individual poachers as well as organised gangs.

One Van Gujjar commented that "the only way to protect the forest is to have an honest police force patrolling 24 hours a day." As noted previously Van Gujjars indicated that they were either ignored or punished when they reported poaching and have thus adopted a strategy of ignoring it. They note that poachers typically usually avoid the immediate forest around their deras and this provides some refuge for animals. A Ranipur Van Gujjar also noted that poachers tended to avoid areas in the daytime where there were large numbers of villagers collecting firewood. Van Gujjars generally report being fearful of poachers themselves, which is the major reservation they have when discussing the possibility of community forest management.

Although the park authorities acknowledge that poaching is a problem, it receives only two paragraphs of documentation in the park plan, compared to chapters on local people and their impact on the forest. The report merely notes that "there is a lack of alertness and vigilance among the staff" with regards to poaching (Kumar, 1995, vol. 1, p. 172). The yearly poaching cases for 1987-88 through to 1993-94 indicate a decreasing trend from a high of 55 in 1987-88 to a low of 8 in 1993-94. Theoretically, one would expect that lower officials would want to inform higher officials, if they were concerned about poaching and wanted more support for taking action. Instead, they appear to downplay it, which supports the Van Gujjar view that they condone it or are benefitting from it. The park plan also downplays it in the overall objectives and obstacles section of the plan, mentioning only tiger and elephant poaching, the latter has not been a problem. The only strategies noted to address poaching involve requesting more enforcement staff and equipment, yet local people do not see this as the way to solve the problem.

It is interesting to note that the District Forest Officer (DFO) in the Haridwar Forest Division, which is outside the proposed park but bordering on it, recently pressed charges against a gang of poachers including the son of a member of parliament. Van Gujjars reported that this same gang had been operating within the proposed park. After pressing charges, the Haridwar DFO was transferred from her position, indicating that the resistance to addressing poaching is a widespread systemic problem so that even courageous people within the system are punished for taking action. Thus, forest officials may be unable to deal with the problem given its systemic roots.

The impact of poaching is also evident in comparing this area with other jurisdictions. For example, the Ministry of Environment and Forests (MoEF) notes that the population of elephants in India has stabilised over the past two decades, whereas it has diminished by 50% in Africa. This is despite the fact that India has a human population density of 250 people per square mile while in Africa it is 25 per square mile. The MoEF attributes the relative improvement in India to the cultural and religious significance of elephants so that it is morally reprehensible to kill them (John, 1995). Similarly, one member of the research team recently visited a rainforest wildlife sanctuary in Thailand which had never been cut and had very dense, undisturbed vegetation with no grazing disturbance. There were few birds however and no sightings of mammals, including monkeys, even at peak activity times such as sunset. Local officials reported that this was due to the traditional Thai practice of hunting a wide range of species including monkeys and songbirds. It provides a sharp contrast to the large numbers of monkeys and birds along the roads in the Rajaji area, despite significant forest deterioration.

(b). Forest Management Practices: The history of forest management practices, regarding reduction in the quality of the forest and the amount of ground vegetation, has already been described, but the practices are also an important factor in reducing the quality and amount of habitat. For example, the development of commercial plantations reduces the food supply for wildlife as well as livestock.

Park authorities propose a number of physical habitat modifications, such as digging wells and restricting domestic livestock use with fencing. There is little point in spending money on additional water holes if poaching is the principal problem. The animals may be shot as they gather around the water holes. Moreover, restricting Van Gujjars from an area reduces the human presence which may increase poaching. Van Gujjars report that Forest Departments, action of filling wells and water holes in the summer is an asset to wildlife. But if habitat modification is desirable as a secondary strategy, it would be wiser and less costly to draw on the knowledge, experience, and support of the people living in the area.

Villagers in the Chilla area expressed distress with respect to some Forest Department actions towards elephants. They noted that in the early 1980s, just prior to the notification of the proposed park, that wildlife authorities captured approximately ten young elephants, which were then sold to foreign zoos for a profit. They viewed this as a questionable practice, in that shortly thereafter the forest officials restricted villager access to the forest to protect the elephant herd. They also feel that the loss of their young increased the adult elephants' perception of humans as a threat and this increased their danger to local people. They also believe that elephants equipped with radio collars for research purposes have become angrier and more dangerous. This may not be a reason for discontinuing the research, but it suggests that local people should be consulted about research practices.

(c) Van Gujjar Impact: It seems that the Van Gujjar presence has a positive impact on wildlife, with the exception of the possible competition for food in the summer season with static livestock. Van Gujjars indicate that even this is not a problem in that herbivores are more flexible in their food sources and can reach locales that buffaloes cannot. In the winter season buffaloes primarily eat lopped leaves. There are inevitably leaves left at the base of trees on a regular basis, and these become an additional source of food for herbivores which they normally could not reach. Van Gujjars report that deer join buffaloes at night and in some cases eat leaves with the herd. In addition, Van Gujjars' deras seem to provide some immediate safety for herbivores and Van Gujjars' presence seems to discourage poaching in the immediate area.

In numbers of instances, Van Gujjars have dug wells and provided an additional source of water in the summer season. The park officials report that Van Gujjars barricade water holes so that wildlife cannot use them. Van Gujjars deny this charge, and note that animals utilise the water holes during the night when Van Gujjars are not using them. These water holes are most important in the summer season, and thus diminished Van Gujjar presence in the summer could reduce the possibility of competition. On the other hand, reduced Van Gujjar presence might well encourage more poaching. Van

Gujjars also note that wild animals move over considerable distances, and thus are quite capable of finding suitable water sources. It is also quite straightforward to dig additional water holes if this is found to be a local issue.

Van Gujjar perspectives on the behaviour of herbivores are supported by the existing wildlife research on habitat preferences of deer populations. Bhatt (1993) studied habitat usage patterns of chital in the proposed Rajaji Park. He noted that chital were particularly likely to be sighted on winter mornings in mixed forest where they were consuming fallen leaves from lopping. He also noted the proximity of a water hole dug by Van Gujjars for livestock as an attraction. Only this tract of mixed forest was used by chital in preference to other tracts which were dominated by teak. Teak is not useful as fodder for chital or buffalo.

A. Clark, H. Sewill, & R. Watts note that "the Gujjars are commonly believed, by the Forest and Wildlife Authorities, to be a serious threat to the wildlife. In the study of wildlife presence the team found that the deer species were not disturbed by the Gujjar activities. Indeed, the deer were reported to feed alongside the buffalo at night and were seen close to dera both day and night. The fodder lopped by the Gujjars and left on the ground provides the deer with an additional food source. The study of the larger carnivores was not undertaken. The Gujjars lose a few buffalo calves a year to the large carnivores. However, during the interviews the Gujjars seemed quite philosophical about losses and did not complain about competition with any wild animal."

Bhatnagar (1991) studied habitat usage of sambar. This study showed no difference from lopped to unlopped areas but did indicate a preference for hill areas (where Van Gujjars reside) over plains areas (border zones of the proposed park subject to grazing of village livestock). Disturbance reduced sambar presence in the plains, but not in the hills. The author suggests this is due to differential effects of disturbance on the ground vegetation. In the plains, disturbance (grazing) resulted in a high proportion of weeds which are not palatable to sambar dominating the ground cover. Disturbance in the hills in the form of lopping opened up the canopy and encouraged undergrowth, but in this instance the additional undergrowth was dominated by shrubs which are palatable to sambar. Scientific research again seems to support indigenous knowledge.

Van Gujjars seem to have co-existed with elephants over generations, and there is little evident competition. A study of the feeding habits of a radio collared elephant in the proposed Rajaji park found that 60%-70% of feeding was on bark with the remaining portion equally divided between leaves and branches of selected species (Sale et. al., 1989). Elephants were rarely observed to eat grass. Thus, there appears to be little overlap between buffaloes and elephants with respect to food sources. Moreover, where they eat the same species, lopping results in excess leaves on the ground which is of benefit to the elephants. Elephants travel across large areas and are not blocked by the dispersed Van Gujjar population. They also move to water sources, and Van Gujjar wells provide additional sources for them. Despite the fact that elephants can pose a threat to Van Gujjars by occasionally invading deras at night in search of oilcakes, Van Gujjars accept the damage and danger as an element of their lifestyle, recognising the elephants' rights in the forest as well as their own. An elderly woman from Rasulpur Khol commented that "neither the elephants nor the Van Gujjars came from away. We were both born here and we live together."

A recent incident involving the rescue of a baby elephant shows the different approaches to interacting with wildlife between forest officials and Van Gujjars (The Hindu, 1996). A baby elephant fell into a well and was trapped while the nearby parents were unable to help. Forest officials approached the area and fired shots in the air to scare off the parents, but the parents charged the officials who fled. In contrast, Van Gujjars waited for the night, by which time the parents were away from the scene. They then lifted the elephant out of the well, nursed it in front of a fire and offered some warm milk. Later the parents returned and the baby rejoined them. Their indigenous knowledge and their continued presence in the area allowed them to intervene effectively. Another Van Gujjar reported that a hurt elephant spent nearly a week near his dera along with the buffalo herd, and was nursed back to health before returning to the forest.

Finally, Van Gujjars' presence is also helpful to carnivores because it discourages poaching, although carnivores usually spend most of their time somewhat away from the deras. Carnivores will also thrive with an increase in their herbivore food supply, which again requires a reduction in poaching. Van Gujjars could be even more helpful if they were encouraged to play a positive role in reporting poaching. This is unlikely under the existing forest management structure, given the history of Van Gujjars being punished for reporting poachers.

(d). *Industrial and Urban Encroachment:* Encroachment by the army, the Chilla Power Canal, and government resettlement of Tehri Dam evacuees pose a significant threat to the elephant population in that they block the only viable elephant migration corridor between the eastern and western portions of the proposed park. If elephants do not move through the corridor then the herd becomes segmented with a loss of genetic diversity that is crucial to a healthy population (Johnsingh et. al., 1990). Local villagers also report an increase in crop damage due to elephants in recent years, and feel that this relates to the inability of elephants to move along their traditional migration routes.

The power canal on the east bank of the Ganges was constructed in the mid-1970s, and elephants are not capable of traversing the steep slopes, and are unwilling to utilise the crossings now available, either a tunnel or one of several narrow bridges. The army munitions dump as well as the Tehri Dam evacuees lands are directly in the elephant path on the west bank of the Ganges. The army has not relocated the ammunitions dump despite a series of negotiations in the late 1980s about alternative sights. Tehri Dam evacuees have not opposed moving if they are given what they feel is reasonable compensation, but again there has been no real progress despite some discussions over the years. There has been no follow-up on the mid 1980s proposals to build bridges over the power canal. In short, the encroachments pose a several threat to the elephant population, which is the lead conservation priority of the proposed park. Park authorities however, have but taken appropriate protective measures over the past decade despite repeatedly identifying the problems. The Haridwar district headquarters and the expansion of a number of temples in the park also demonstrate a failure to protect the ecological integrity of the park area effectively in the face of encroachments.

(e) Villager Impact: The level of villager grazing has already been identified as a significant pressure in the depletion of ground vegetation. This factor appears to have some impact on herbivores such as sambar as well. Research shows that sambar prefer relatively undisturbed areas in the plains, though the research study points out that it was impossible to find completely undisturbed areas (Bhatnagar, 1991). Again it seems that it is the magnitude of the grazing, the failure to improve the forest for fodder, and the lack of villager responsibility for the forest, that constitutes the problem to wildlife, rather than the existence of grazing. A particular area of difficulty involves grazing in the elephant corridor. If the encroachments were ever to be removed, it would be essential for elephants to have fodder in the corridor in order to use it (Johnsingh et. al., 1990).

Both villagers and Van Gujjars report that some poachers and illegal fellers actually live in the local villages, though they represent a small criminal element. Villagers, like Van Gujjars, indicate having been ignored or punished for reporting illegal activities. Moreover, villagers view the park authorities as having the responsibility for wildlife, and given the loss of their traditional rights due to the proposed park, they are unwilling to go out of their way to help park officials do their jobs. Wildlife also threaten their livelihood through crop predation, and forest officials have not responded to their requests for compensation (Pottie, 1995). Although the vast majority of villagers disapprove of the illegal activities, they are not presently acting to curb these activities. They may discourage poaching in areas where there are high levels of daytime firewood collection by their presence in the area.

Synthesis of Factors in Wildlife Problems : Poaching appears to be the major threat to wildlife, and the present enforcement system is not able to control it. Van Gujjars and villagers are a potential resource through which to address the problem, but their treatment by forest officials is such that they are unwilling to act. Van Gujjars have a positive influence as the result of having an established residential presence and there would likely be more poaching if they were not in the forest.

Van Gujjars do not appear to have a significant impact on wildlife through their livestock and lifestyle. Their deras and indigenous knowledge are helpful in periodically protecting wildlife. The amount of villager grazing pressure does appear to compete with herbivores, and there is no effective system to regulate it or encourage villager responsibility for the resource. The principal threats to elephants are the encroachments in the elephant corridor. Overall, the present management system has been unable to counter the threats to wildlife.

Increased Rate of Soil Erosion

Status

Erosion is and has been high due to the geology and slopes of the Shiwalik Hills. Park authorities go further in suggesting that there has been an increase in the rate of soil erosion in recent years (Kumar, 1995, vol. 1, p. 63) but no evidence is cited to support this contention. The report presents data on the width of raos across decades starting in the 1920s (ibid., p. 215). Analysing this data for the rate of increase across 20 year spans, we note that there has been in fact a decrease in the average rate at which the raos have been widening over years (see Table 10). Moreover, there is a significant amount of variability among raos. The official view that soil erosion is an increasing problem is invalid, though it remains a serious problem at its present rate.

Analysis of Contributing Factors

(a) *Forest Management Practices:* Clear felling and over-exploitation are major causes of soil erosion. The end of commercial forestry operations and the initiation of the proposed park has had a beneficial effect. The Forest Department has erected physical erosion barriers in selected sites across the proposed park over the years which may also have had some positive impact. Villagers report that the removal of stones from the raos by contractors in the Mohand range contributes to erosion there.

(b) Van Gujjar Impact: Van Gujjars report that lopping increases undergrowth and reduces soil erosion. Our field observations and the available research studies of Van Gujjar areas support this contention (Bhatnagar, 1991; Clark et. al., 1986). Bare ground, which increases soil erosion, is not associated with lopping in Van Gujjar areas. Clark et. al. (1986) did find evidence of erosion in a small number of steep survey sites in Van Gujjar areas (15% of sites) due to bare ground caused by animal paths.

(c) Villager Impact: Over grazing in boundary areas may result in an increase in the amount of bare ground and soil erosion. The associated spread of weeds however, may at the same time provide additional ground cover in these areas, and counteract the negative grazing impact on soil erosion. In addition, grazing pressure is highest in border areas, which in turn are flatter and less prone to erosion.

(d) Synthesis of Factors in Soil Erosion: It is unclear whether geological and/or climatic factors are responsible for the decreasing rate of soil erosion, or human intervention or interference. The halt in commercial forestry in recent years is likely making a positive contribution, while Van Gujjar presence may be having a positive impact through lopping. Soil erosion remains a serious problem given the Shiwalik terrain, climate and soils.

Table 10
Width of Raos at Shakumbari Road in the Proposed Rajaji National Park

| | Width of Raos (meters) | | | | Period Differences (meters) | | | |
	1928	1948	1967	1987	1928-48	1948-67	1967-87	Total Diff
Rao								
Mohand	191	262	330	325	71	68	-5	134
Sukh	135	167	170	210	32	3	40	75
Chillawali	213	238	336	342	25	98	6	129
Gaj	76	176	188	184	100	12	-4	108
Andheri	58	56	93	124	-2	37	31	66
Beenj	109	126	183	155	17	57	-28	46
Dholkhand	185	146	160	162	-39	14	2	-23
Malowali	69	109	93	72	40	-16	-21	3
Bam	64	106	107	132	42	1	25	68
Sendhali	46	69	8	42	23	-61	34	-4
Betban	362	385	366	408	23	-19	42	46
Gholana	80	91	107	132	11	16	25	52
Avg. Diff.					28.58	17.50	12.25	58.33

Data excerpted from Kumar (1995)- *Management Plan of Rajaji National Park, Vol. I,* p. 215.

Forest Fires

Status

Fires are described as a major problem, regarding destruction of ground cover and deterioration in ground litter, according to the park authorities. The number of reported fires from 1987 to 1993 ranged from a yearly high of 20 (144 hectares affected) to a low of 4 (19 hectares affected). However, both the park plan and Van Gujjars we interviewed indicate that forest officials do not report all fires and also underestimate the damage incurred. The actual magnitude of problems due to fire is unknown.

Analysis of Contributing Factors

(a). *Illegal Activities:* A wide range of Van Gujjars indicated that fires are most frequently caused by persons involved in illegal activities. Poachers were reported to set fires in order to herd wildlife; illegal fellers use fire as a tool to hide their activities; and other people burn grass in order to uncover horns.

(b). Forest Management Practices: Several Van Gujjars noted that forest officials are often slow to respond or never respond to initial fire reports. A Van Gujjar in Kansrao cited an instance when he reported a fire and the response was slow, and involving the forest official coming to investigate the scene. Only after the investigation was there any attempt to call for assistance to fight the fire. Both the park plan and Van Gujjars note that the fire lines are poorly maintained.

(c). Van Gujjar Impact: Van Gujjars reported that historically they would mobilise fire fighting on a large scale. Their increasing insecurity however, with respect to land tenure, slow responses and lack of support from forest officials, has resulted in a more recent tendency to let the fires burn unless they directly threaten their forests. Authorities accuse Van Gujjars of causing fires in the winter, but provide no evidence to support that claim. In fact there are no statistics presented on the cause of fires, and we saw no evidence of a winter fire problem during our field visits over this period.

(d). Villager Impact: Villagers also indicate that they are less willing to fight fires than in past times, given the attitude of forest officials and their poor relationship with the Forest Department. They point out that harvesting bhabbar grass is a deterrent to the spread of fires.

(e). Synthesis of Factors in Forest Fires: Illicit activities appear to be the most prominent human factor in inducing forest fires, but there is little presence information on frequency and causes. Van Gujjars and villagers could make a greater contribution to fighting fires than at present if they felt more responsibility for the wider area, and/or saw a meaningful commitment from forest officials to prevent and fight fires.

Conclusions : Ecological Problems and Human Impact

It is evident that there are serious ecological problems with respect to forest deterioration, depletion in ground cover, weed infestation, wildlife problems, soil erosion, and forest fires. The official forest plan provides a range of information on these problems, but then primarily attributes the negative impacts to Van Gujjars, local peoples, and insufficient park resources. A detailed analysis of the official information, the findings provided by external empirical research studies, our field observations, and the perspectives of a wide range of Van Gujjars and villagers, invalidates this official park view.

This report has documented past destructive forest management practices, present poaching and tree felling with the support of some officials, and the extensive system of illicit fees resulting in over- exploitation by local people, as the principle causes of the ecological problems. A major implicit factor governing forest management is the use of the forests to generate income for local officials. The overall incentive structure is such that all involved are caught in a system which requires individuals to act for personal gain rather than forest protection, regardless of their individual or cultural values. It is extremely difficult, if not impossible, for an individual to fight this system.

The ecological problems of the proposed Rajaji Park are due to overall structural problems in the management system. A wide range of evidence indicates that the present system is failing to address the major problems, and yet the authorities recommend strategies which are amplifications of the present approaches. They propose to:

- increase the park area, although the system is not effectively managing the present area.
- further restrict the traditional rights of local people without consulting. Yet this would lead local peoples to ignore the restrictions. Restrictions would also be ineffective in that they could be used as a pretext to extract more illicit fees, resulting in greater resource exploitation.

- increase field and enforcement staff. Yet this would result in larger numbers of people working within a management system, in which there are not enough incentives for staff to act for protection.

- obtain 130 lakh funds for one village for "participatory" ecodevelopment to be implemented through a broad rural development process in order to reduce the village's dependence on park resources. Yet centrally funded rural development funds in India have a poor record in reaching the grass roots beneficiaries. There is no provision for increasing villagers' responsibility for the forest they presently use within the proposed park boundary. Finally, the possibility of obtaining this level of funding for more than 100 villages around the proposed park is negligible.

- allot large amounts of money (i.e., 79 lakhs) for physical development projects (i.e., erosion barriers) covering a small area (2 blocks). Yet soil erosion is not the foremost problem, and the likelihood of implementing such a costly approach throughout the area is remote.

- remove Van Gujjars from the area. Yet this would eliminate the group that has the greatest direct interest in preserving and protecting the forest, as a basis for its own economic and cultural survival.

Implementation of this plan would increase the existing problems. The ecological problems of the Rajaji area are, in fact, social systems problems that require a restructuring of the management system. Decision-making must be placed in the hands of groups with the highest investment in forest protection. Behavioural incentives must be structured so as to encourage behaviour that reduces human impact on the environment. The management process must be open, monitorable, and accountable to avoid any one group from exploiting the forest for its own benefit.

In short, the ecological problems of the Rajaji area are one example of the failure of international and national policies for protected areas which were described in part I of this report. The foregoing detailed analysis of the broader policy context and alternatives points to Community Forest Management in Protected Areas (CFM-PA) as holding the most promise, assuming that local conditions and community characteristics are those which have achieved success in other areas. The next chapter examines these conditions and characteristics to demonstrate that CFM-PA is the most appropriate strategy for the Rajaji context.

PART - THREE
LOCAL PEOPLE AND PROBLEMS

Part I of this report reviewed the failure of international and national policies of the past and present, along with the widespread conflicts between local peoples and protected areas which have resulted from them. These policies have failed to address the needs of the environment and the needs of local residents. A review of alternative approaches indicates that local people's involvement is essential, and that community management strategies have been successful in a number of situations in protecting both the ecology and the lifestyle of local people. Part II examined the problems of the Rajaji area from the ecological perspective, indicating that the problems are severe, and the present system cannot solve them. Part III examines the Rajaji context in relation to the characteristics, values, and needs of local people. Different groups are dependent on the park area, and it is important to recognise each group's traditional rights, needs, and potential role in a community approach. Members from each group have been interviewed with respect to their perspectives, needs, and potential solutions. This section however, emphasises the cultural background and problems of the Van Gujjars because they are the only local people living in the core areas of the proposed park, and would be the lead managers of the core area. Their unique forest-dwelling lifestyle is essential to effective community forest management in the Rajaji area. They have also been the primary group targeted for eviction by the park authorities in the past. This section also identifies the problems and needs of the Taungyas and local villagers as they are proposed to take on lead management roles and responsibilities for the resources and boundary areas where they are the primary users.

Chapter 6
Culture and Ecological Values of the Van Gujjars

In the following chapter we analyse what it is in the personality and culture of Van Gujjars which would make them responsible custodians of the proposed park. To this end we dissect their current practices in relation to the forest and with the animals there, particularly their ways of thinking about these things, including their values, motivations and religious beliefs. Moreover, we have specifically examined the various sanctions which impel responsible behaviour, from the external pressures of law and community to the inner taboos of morality and religion.

Van Gujjars, though fearful of the widespread evil in the outside world as they have experienced it, are largely indifferent to the opinions of others and have little impact on their sense of self. Yet they are intensely responsive to public opinion within their own community, since they are strongly motivated by their need to gain public honour and respect. Thus we have analysed in some detail the behaviour which brings credit to a family and their lineage, and what brings disgrace. For example, a prospering forest was found to bring honour, not only in this life but the next, whereas abuse of nature is believed to be a curse on God, bringing personal shame and destruction to generations yet unborn.

One can distinguish within Van Gujjar society a series of sanctions acting in parallel to ensure the preservation of their forest home, from self-interest and the desire for respect within their own close-knit and proudly independent society, to religious sanctions and those secular punishments meted out by the panchayat, including public humiliation, ousting and even death. The result is a system of values highly favourable to ecological responsibility. This chapter is organised according to the prominent cultural themes and values which govern the Van Gujjar' lifestyle in the Rajaji area.

A Self-Contained Society

The Van Gujjars in and around the proposed Rajaji National Park comprise a branch of the larger Gujjar tradition of pastoral nomads with representatives from Afghanistan and Pakistan, through Kashmir, Jammu and Himachal Pradesh, to Garhwal and Northern Uttar Pradesh. (Binghay, 1899; Khatana, 1992; Munshi, 1955; Rawat, 1993b; Shashi, 1979).

The Van Gujjars have inhabited the Himalayan region as nomads for centuries, but their transient lifestyle has moved in relation to social and ecological pressures exerted on them by more powerful forces. Thus, their traditional wide range of movement has diminished over generations. For pastoral nomads, home is defined by their ecological setting and yearly movements. It is difficult to ascertain when they first entered the Rajaji ,area but Williams (1874, p. 29) notes Van Gujjars coming to the Doon Valley in the 18th century, and describes the exploits of a band of Van Gujjars who were viewed as outlaws by the British in the 1820s. Letters among forestry officials in the late 19th century also indicate that Van Gujjars had been in the area for a very long time (Gooch, 1994). By the late 19th century their free movement was already curtailed as the British had set up check points along their migration routes to control the numbers of buffalo passing through (Gooch, 1992).

The Van (or forest) Gujjars have traditionally been buffalo herders, making their home, during the winter season among the dry deciduous forests of the lower Shiwalik Hills, some 1,000 to 2,000 ft. above sea level, then travelling on foot with all their possessions 200 to 300 kilometres to the high mountain pastures (at 8,000 to 12,000 feet) for summer grazing. Unlike some tribal communities in which use of land resources is wholly shared, Van Gujjars have evolved a tradition in both their winter and summer pastures of having separate areas of forest allotted to each family, often a joint one. Thus, a Van Gujjar family knows exactly which trees it can use and which ones belong to the neighbours.

Their summer pastures are widely scattered; those families to the West of the proposed Park tend to migrate to the Simla area of Himachal Pradesh and to the Western edge of Uttar Pradesh, whereas more Easterly families follow the headwaters of the Ganges into the hills above Uttarkashi or into the Chamoli District. Some families move shorter distances to local sources of water in the summer season.

Travelling mostly at night along narrow mountain roads heavily used by buses and trucks, the Van Gujjars travel in family groups, the buffalo in the lead, led by the oldest female. It is said the buffalo themselves initiate the migration, becoming restless for the hills as the April heat increases, and then becoming anxious to descend as the mountain air cools in late September. They know the regular stops along the route and where water is to be found, though in recent years increasing pressure from hill landlords has meant more and more restrictions, so that forage has to be purchased at high prices and fees paid for access to water.

The men follow the buffalo, staff across shoulders, shielding the buffalo from danger by walking directly in front of passing trucks, since they can be more readily seen than the black buffaloes in the darkness of the night. Even then, buffalo are not infrequently killed or injured, and a major incentive to literacy (together with the need to keep a check on moneylenders) has been to note down the licence numbers of trucks before they can take off.

Close behind come the few cows and goats most families possess, driven by the women and children, along with some ponies or oxen to carry their few material possessions (pots and pans, sickles, digging sticks, or black plastic sheeting for temporary shelter along the way). In recent years, a few families have sent some of their possessions or animals by truck over the more difficult stages of the journey, but not many can afford such a luxury.

Passing through the villages en route, the Van Gujjars remain detached, distinctive in their bearing, dress and physical appearance. In market towns such as Mohand, just to the West of the proposed Park, visiting Van Gujjars are

135

equally distinct with their dignified detachment and steady gaze. Frequently described as tall (e.g., Khatana, 1992; Iqbal & Nirash, 1978), they are, in fact, not particularly, males averaging approximately 5ft. 7in. They convey the impression of height however, from their lithe body-build and erect posture, derived perhaps, from their practice of walking with both arms resting on a staff thrust across the shoulders and behind the neck. Moreover, many of their most prominent male leaders are exceptionally tall as stature (i.e., "presence") is an important factor in the making of a respected man in Van Gujjar society.

Physically, Van Gujjars resemble many of the hill tribes of Afghanistan and Jammu-Kashmir (from where they insist they come; Lidhoo, 1988; Dupree, 1980), having rather gaunt faces with deep-set eyes, large often hooked noses and full lips. Relatively light skinned with an abundance of dark body hair, their facial hair tends to turn white at a relatively early age (40-50 years), a fact used to advantage by the older men who dye their well-trimmed beards bright orange with henna.

A full beard is felt to be essential to a Van Gujjar male, though a moustache is optional, often being reduced to a thin line, though some twist the ends in the Kashmiri style. Older Van Gujjar men will swear by their beards, for in their beards their dignity is said to lie, and it is the greatest insult to pull another's beard. (Oaths may also be sworn standing in water or holding a vessel of Ganges water.)

The shaven head is always covered, either with a turban (Gj. 'pug') or red pointed cap (Gj. 'topi') embroidered with floral motifs in pastel pinks. Such caps, which may be found in modified form in Kashmir, are now more often worn by children of both sexes than adult males. A simple wrap-around cloth (Gj. 'loongi') covers the legs almost to the ground, while a loose-fitting shirt or 'Kurta' covers the upper body, over which a 'jawaharcut' or waistcoat with many pockets may be worn from which an amazing variety of objects may be extracted. A woollen blanket, often patterned with bright red checks , completes the attire in cooler weather.

Unlike the prevalent surrounding custom, Van Gujjar men greet each other with firm handshake and steady gaze, eye to eye, often with a slight gesture towards the heart. If they are close friends or kin, they will gravely embrace three times; once heart to heart, then right side against right side, then heart to heart again, with a muttered; "Salaam Wale Kum..... Wale Kum as Salaam." (Women, or men greeting women, usually embrace only once.)

Whereas public life is the domain of the male, a woman's place is largely in the dera or home, so women are less evident at market and public gatherings, and their clothing generally tends to be more subdued. Saris are never worn but a loose fitting "suit" with shirt (kameez) and baggy pants (salwar) taken, in at the ankles in the Punjabi style. The head is always covered, usually by a simple 'chadar' or sheet, folded in a variety of ways according to need. There is no purdah or veil and women have considerably more freedom than in many other rural societies, both socially and in their right to own personal property. Women maintain strong connections with their original family after marriage, and play important roles in maintaining accounts and financial transactions with regard to milk production and family needs. Traditionally, the eldest male is the leader or "lambardar" of a dera, but in his absence or death, the eldest woman rather than the eldest son frequently becomes the leader. Older women always play important and vocal roles in family decisions.

Both men and women milk the buffaloes and cut grass, though cooking and preparing milk products is largely woman's work. On the other hand, unlike in many other parts of Garhwal where women also do most of the lopping and farm labour, this is mainly the task of younger males. Both sexes work on dera (i.e. house) construction, the males building the frame while the women cut thatch and plaster walls. Decorations are usually the work of women, but material possessions are generally undecorated and there is a lack of art of any kind. Percussion or other kinds of musical instruments are all but absent (which precludes the kind of ritualised possession states and trance dancing prevalent in many Garhwali villages.). Some maintain that music is forbidden by the scriptures, but the more traditional explanation is that drums are associated with the hunt, and music would scare away wild animals. More poetically, they prefer silence, for, "in silence you can hear the sounds of Nature/God (Kudrat)."

Once story-telling was perhaps a developed art, but if so the old stories, and with them the poetic metaphors, have been largely forgotten. Both men and women however sing or rather chant, the most popular verses today deriving from a book possessed by many of the wealthier families, written in Urdu, and relating the "truths" or "revelations" of Joseph as revealed in the Koran (Taf sir-e-Yusuf). This text is also popular in Pakistan today and may frequently be heard broadcast, but its use among the Van Gujjars seems to stem from the fact that it is one of the few texts available to them in their dialect and is taught by the Holy Men. Other songs, some contemporary, others apparently old, are in their own language of Gugeru (Hindi; Gujari) Such songs are made up of apparently unconnected verses strung together-about current affairs, homely advice, jokes, and fragments whose significance has been lost. (Gugeru may originally have been a Jammu-Kashmiri dialect with a mixture of West Punjabi and Urdu words, pronounced with an altered dialectical stress, though some interpret the matter differently. Today it is much mixed with Hindi.)

Rising well before dawn, women rekindle the fire, distribute milky-tea and begin the daily chores, while the young men scatter to collect the buffalo from the forest where they have spent the night. Buffaloes are milked only once, and the morning meal is taken at this time, with women and children drinking directly from the teat. Most men, however, seem to prefer milky-tea, often with added salt, sipped by the fire.

The main meal of roti and dal is served about mid-day followed by rest, but also an opportunity for talk and visiting. In late afternoon the men call the buffalo to where they will lop trees for fodder, scattering the branches so that the buffalo can feed. Dinner is much like lunch, with more roti (chapatti) and salt dal, occasionally supplemented with vegetables purchased from the market, such as potatoes or okra. Few, if any, jungle products supplement the monotonous diet, except the occasional Jamun fruit (Syzygium cuminii), the small yellowish fruit of the Ber (Zizyphus jujube), or the white flowers of the Khirini tree (Manilkara hexandra). Very few Van Gujjars grow even the most basic vegetables, and indeed those within the proposed park are forbidden to do so.

Many established deras include married sons and, with ample hands to do the basic work, the older men are free to engage in public life. Even in the hills, where deras may be as far as ten or fifteen kilometres apart, a network of trails connects the homesteads. In winter, visiting is even easier and in the forest half a dozen deras may be found within a radius of one or two kilometres. Visitors regularly stay overnight and senior males travel extensively to exchange gossip, settle disputes, arrange marriages and make deals of one kind or another. Women's sojourns are most frequently to visit relations, or to meet the needs of children (e.g., health problems). Gossip is not idle but includes detailed discussions about other deras, the condition of their buffalo and jungle, their hospitality or absence thereof, and so on, for on such details rests the reputation of the dera.

Although there are some differences in Van Gujjar' needs and perspectives based on the size of their herds, in general disparities are small and subjugated to their dominant cultural identity and sense of community. A few Van Gujjars have large herds and more affluence and some of these have grown dependent on forest department favours and manipulation for their continued prosperity. This has caused some internal conflict, but is seen as a problem by most Van Gujjars only in relation to the inability of the community to counter manipulative Forest Department actions.

Mainstream commentators often view the Van Gujjars as a remote culture which has maintained its integrity simply because it has only recently come in contact with mainstream society. As pastoralists they are entirely dependent in fact, on outsiders for trade and the distribution of their milk and their nomadic movements put them in direct contact with a wide range of local people. Pragmatic and down-to-earth in their assessment of life, they are all too familiar with the evils of the outside world with its violence and graft, and they see no virtue in society. By and large a self-contained society, they choose to live in their own sphere of praise and blame, indifferent to the opinions of outsides. Time and again we were told; "We do not want to become like other men......all we desire is to be left alone to live our lives free from theft, anger and corruption." Or again; "The forest is like a veil behind which we can live our lives...... do not take this veil from us!"

Earning Respect and Honour

Each dera, with its surrounding territory (jhera), is a continually self-perpetuating unit, with its own history and reputation to uphold, nurture and develop. To this reputation every member may contribute, but it is chiefly embodied in the dera 's leader, usually the eldest male, the Lambardar. Any offence against the dera or desecration of its surroundings is a personal slight and any praise adds to his reputation and fame.

When asked what brings honour to a person, hospitality is usually mentioned first, followed by generosity, for, "If we were not hospitable we would not be human beings" (Gj; insaan). Protection of the weak and poor is highly valued, as are personal restraint and the ability to settle disputes. An older male should have the ability to speak gravely and with conviction in the general assembly (Panchayat), though justice is said to be more important than flowery speech. Age itself is grounds for respect, though it should be accompanied by wisdom and personal dignity.

Reputation builds slowly over the lifetime of the individual, but it is easily lost through disgraceful conduct such as lying, cheating, stealing, fighting, flirting with another's wife, or disrespect for elders. At first, because it seems inconceivable, the maltreatment of animals or the forest is seldom mentioned. And in fact, many deny that it could ever happen. Van Gujjars however, are sure that if it did, it would bring immediate disgrace, and like the other 'sins' mentioned above, would be grounds for public condemnation by the Panchayat, possibly even beating and ostracism. By contrast, well-fed buffaloes and a healthy forest give honour to a man. Nature, which is also God, smiles on such a one, trees bloom, the buffalo bring forth milk and the good man prospers. "If we keep our forest well, we gain respect in our community just as the Prime Minister does in his country."

Buffalo, the Van Gujjar's Livelihood

Van Gujjar buffalo are not the rather dopey animals one sees commonly in village India, but a livelier and altogether more robust breed with the endurance to cover great distances on very little food, and the agility to scramble over rocks in high mountain pastures. Sleeping quietly with their eyes closed or wallowing contentedly in a mud-hole, they come alive after dark and are said to possess excellent night vision. Either sex is quite capable of fending off predators, and indeed deer are said to seek out their protection. Easily distinguished by appearance and personality, each beast is named individually, just like humans.

Buffalo are a family's pride and are treated as members of the extended family. They are constantly handled and massaged during milking, and are even fed some of their own milk at this time, either directly from the teat or if they decline to turn the head, then by squirting it into the open mouth. Each buffalo is treated individually in relation to its needs for food and habitat. Milking mothers get special supplies of rich fodder leaves and herbal medicines immediately after giving birth. Injured animals are carefully tended, and unproductive animals are neither killed nor sold, for, "they have fed us as our mothers with their milk, so why should we not be kind to them?" Injured legs are carefully bound with splints, after which the injury will either heal in 10 to 15 days or the animal will die "naturally" from being unable to forage. Young males are neither sold or killed, but most die naturally as they are given little food. Dead animals are never skinned, for, "How can you skin your mother?" Rather they are buried whole, often with a prayer similar to that used for human burial.

Even a buffalo which does not become pregnant is respected as lucky, in that it makes one appreciate the gifts of Nature.

Although they are treated as family, Van Gujjars are well aware that buffalo are not actual family but are both their care and their livelihood. No kin terms are used when addressing them, and they are not invited to weddings or other festivals, even in a symbolic way. Nor are·they considered sacred as cows are by Hindus. They are treated however, with great affection and respect, and are never eaten.

Forest Creatures Versus Other Animals

Besides buffalo, most Van Gujjar families keep cows, a few goats, and several horses or bullocks for carrying loads during migration. Not all animals however are regarded alike, and distinction is made between the animals of the forest, with whom the Van Gujjars feel a particular affinity, and other animals.

Buffalo, for example, are considered creatures of the forest because they come alive at night, grazing in the open jungle with deer and other herbivores. Such mingling is particularly evident at water holes during the dry season, when there is less food, and when deer gather where lopped branches fall. Thus it is said; "There is no difference between wild animals and our own, for they feed side by side."

All forest animals, whether fierce or gentle, are seen as interrelated components of the natural world and therefore to be cared for by mankind. "They are our children," it is said, "for they have the same life as our own." Moreover, "buffalo have us to care for them, but just because wild animals have nobody to care for them does not mean they should be harmed."

Jackals, snakes, leopards and even the irascible elephant are all given a respected place within the moral order of the natural world. Though different, "like the five fingers of the hand", each has its part to play in the great scheme of things and no vengeance is taken when elephants destroy deras (as they occasionally do) or leopards make off with calves or goats, for such is the way of the natural world, and one animal gives its life for another much as trees sacrifice their leaves to deer and buffalo. Moreover, when an animal's time has come, its death is the will of God.

Since they are regarded as kin, no creature of the forest may be killed, and even injury to one is cause for personal disgrace and discipline before the elders. No forest creature may be eaten or skinned, and if found dead should be buried preferably with a prayer similar to that used for human or buffalo burials. Such practices are further reinforced by the belief that the Koran specifically forbids the skinning or eating of wild animals (though this is not strictly true). Moreover, as is usual, Van Gujjars tend to back ethical practice with pragmatic reasoning, maintaining that if they did not bury the dead, vultures would come and foul the leaves. If buffalo eat such leaves they are believed to go mad. For this there is said to be no cure except readings from the Koran, Hindu mantras or injection with allopathic drugs.

In contrast to buffalo, goats are not creatures of the forest and thus may be eaten and skinned for religious feasts. They "have no unity" in the forest, (i.e., tend to scatter) making them vulnerable to predation and have to be locked in closed quarters at night and herded by children by day. Van Gujjars remain almost completely vegetarian, but as Islam enjoins the eating of goat or sheep at feasts such as that celebrating the sacrifice of Abraham, Van Gujjars may take a little meat on such occasions, while goat skins are used as prayer mats in some households. Additional reasons for keeping goats include the fact that they can be easily sold and thus provide a ready source of income in times of need during migration, or they can be traded for an errant buffalo which has strayed on to a villager's land and is being held. Goat milk is never used for human consumption.

The position of cows and horses is ambiguous. Smaller individuals are brought in at night but larger ones are generally allowed to roam, though "they have little unity". Cow's milk, being less rich than that of buffaloes, is fed to children, particularly if they are sick and to sick buffaloes, providing an additional reason not to eat them, since Van

Gujjars decline to eat the flesh of animals whose milk they drink, for, "how could we harm our mothers?"

Other domestic animals such as chickens and sheep are usually not kept, perhaps because they would too easily fall victim to predators. Jungle fowl occur in the wild and thus chickens could be considered creatures of the forest. Pigs occur naturally in the wild, and though they would make easy prey are never harmed. Moreover, the Koran expressly forbids the eating of pig.

Dogs, usually of the Bhotia breed, are kept as watchdogs, and are treated with considerable affection. Chained and sleep by day, they are fed the same vegetarian diet as other members of the family. Like other forest creatures, they come alive at night when they are permitted to run free. They are strictly trained however, to remain around the dera, and are never permitted to chase or harm wild animals.

Their Forest Home

Van Gujjars return to the same dera sites year after year unless prevented by outside forces, such as villagers or forestry officials. A family comes to know its own territory in intimate detail, noting every seasonal change and variation. Thus, to walk through the forest with a Van Gujjar guide is a lesson in biodiversity; every species of tree is known, its quality as fodder, the timing of its leaf-fall, medicinal properties and so on. Every sound has meaning, every bird known and its habits noted, every fallen branch or tree noted. Van Gujjars readily confess that they grieve in their hearts for dead or dying trees, and complain that villagers have no feeling for trees and so do indiscriminate damage.

Not all trees however, occupy an equal place in the Van Gujjar heart, and distinction is made between species useful to forest animals and those which are not. In times of need, branches from the latter are cut for dera construction, though use may be made of any kind of fallen or deformed wood.

Lopping, the harvesting of leaves for fodder, is an art learned early in adolescence when a boy is strong enough to climb to the top of the tallest tree. Individual trees are largely

The Van Gujjars

A small Van Gujjar dera under construction showing the use of dead wood

The two photos show difference between a lopped Bahera (Terminalia Balerica) tree and a pair of unlopped trees of the same species in the Chilla range. The lopped tree has a narrower canopy with thicker and deeper green leaves. Of the two unlopped trees, the smaller one on the left is the same age as the lopped tree while the larger Bahera is significantly older

A Van Gujjar family on their yearly migration to the mountains

A Van Gujjar lopping a fodder tree

Camels transporting Bhabbar grass out of the forest

A Van Gujjar woman

A Van Gujjar woman skillfully works on a topi - their ethnic identity

A tree in the park which was felled and then stripped of its bark. The
bark would have been sold for use in the production of alcohol

A Van Gujjar points to a dead Neelgai as a result of army firing practices

Guns getting ready for destruction during army firing practices in the forest

A Van Gujjar points to a tree in the Mohand Range which was destroyed by an army artillery shell during firing exercises

A Van Gujjar meeting to discuss the CFM Plan

Van Gujjars learning to read and write under RLEK's literacy programme

The Railway station at Kansrao in the proposed park

A barren forest floor in a tree plantation on the border of the proposed Park.

The Chilla Power Canal

Van Gujjars working together for their Self-Help Milk Marketing Co-operative

stripped, save for the leading shoots and some leaves which are left intact. One large tree provides a day's fodder sufficient for three to five buffalo. Species are lopped in sequence, just before their natural leaf-fall, which can occur from October onwards. By March the forest is almost bare of leaves, and from then until the Spring migration, the buffalo are dependent upon dried grass stored from the previous October and November. Such grass is vital to the survival of buffalo during the spring months and cutting is given top priority by the Van Gujjars on their return from the mountains, taking precedence even over rebuilding deras destroyed by weather and time, villagers, or auctioned off by forest employees. [Theft of structural wood from deras constitutes a major drain on forestry resources and has been amply documented (Bhatia, 1995; Sharma, 1995). Similarly, deserted deras in the hills are 'mined' for timber by competing pastoralists, herb gatherers, etc.]

Realistic in their approach to the forest, one Van Gujjar said quite pragmatically; "The forest is our home, why should we destroy it?" or another; "The forest is our nest......" Thus, they only take what they themselves need, and there is no tradition of selling forest products of any kind, whether wood, honey, rope or medicines.

The Natural World

Down-to-earth in their approach to life, Van Gujjars picture society and the forest in naturalistic rather than symbolic terms. Their deras have no doors and a gap between the walls and thatch permits birds to flit freely in and out. Langur monkeys cluster around the deras at night and jungle sounds fill the air. In warm weather Van Gujjars frequently sleep out of doors and most forest deras have an open-sided sleeping platform as an annexe to the main building, and the evening fire is often placed outside under the stars, for as one old man explained; "An open fire is like a celebration", whereas to gather round an enclosed city fireplace is "like mourning". Given such continuity between the home and natural world, it seems that Van Gujjars should perceive both themselves and the animals as derived from the same forest womb and therefore kin. The forest is a benign and nurturing place, free from the corruption of city life. Nor is it threatening or malign. There are no evil spirits lurking in its depths, and death is a natural fact of life; or, expressed in their terms, an expression of God's will which we may not presume to understand.

According to Mustooq Lambardar, a Van Gujjar of Gohri Range, "There is a harmony (Urdu; *tal-mail*) in our lives, in our forests... amidst these flowers, these wild flowers, and in our animals. We want this harmony to continue... a harmony which has continued for the last 1000 years. Van Gujjars are not criminals (Urdu; *badmash*), they are honest (Urdu; imandar) people. You see this hut of straw... there is no lock, no door... it is very natural (Urdu; kudrati). If we were not natural, criminals would rob us and wild animals would eat our animals... that is natural (Urdu *kudrati*) too. If our animals were kharaab (i.e. devious, impure, bad) then wild animals would eat them. This is a natural thing... that there is a harmony between wild animals and our animals and also with this land, a harmony possible only in the absence of corruption and impurity (Hindi; apavitra). If we leave anything in this house unprotected... it will be preserved... it will be saved... because the people are pure (Hindi; pavitra). And this is natural too, that our animals will not be eaten by wild animals and our children will not be harmed by elephant, tiger or other wild animal... because they, too are pure. Where ever our people live, whether in the forest of the plains or in the mountains... there is a harmony between ourselves, our animals, the forest, the wild animals... this is all related to the herbs of Himalayas... and that too is a natural thing. We don't eat meat, don't gamble, steal... dairying is our only occupation, our livelihood, and by the grace of the Master (Urdu; Malik) we have all we need. Our children are raised in purity... raised on the straight path... and are thus fitted to care for this precious forest."

Thoroughly integrated with the natural world, those who honour Nature's way understand that prosperity exists, without the intervention of anthropomorphized spirits, just as poverty will automatically strike those who scorn her laws.

Nature ('Kudrat') is itself believed to distinguish between the good and natural person and those who are corrupt. Thus, the good person who reveres nature will always be supported by nature (as they are by God), but " a bad person is always suffering... from accidents, sickness... falling out of trees... being attacked by wild animals. To be devoted to Nature (*Kudrat*) is to be devoted to the One above." (Van Gujjar from Bam Kohl, Dholkhand Range). This perspective is well illustrated by Mustooq Lambardar of the Gohri Range, one of their most eloquent leaders.

Sickness, like death, is generally treated in natural terms with few, if any, moral connotations, and modern medicine is readily accepted. Nor is illness generally thought to be caused by offended spirits of the forest nor by irritated ancestral spirits, though if pressed, some admit it might be so. We heard no mention of "dosh" in the sense of moral or transgressional fate, such as is used as an explanation for misfortune by many villagers in the Garhwal.

Some, however, have heard of Jinn and profess that they could be a cause of human trouble if offended, but understanding seems vague. More important in ensuring correct behaviour may be a belief that the ghosts or "powers" (baraka) of righteous ancestors dwell in certain parts of the forest, particularly around Pipal and Banyan trees. There the spirits of the saints have their community. Occasional offerings of sweets or oil are made at such holy sites, though the powers are largely honoured simply by behaving decorously in such places. A story is told how a man once committed wrong doing in such a place, and that very night a leopard took two of his buffalo. Similar beliefs invest certain piles of stones and other ancient structures in the forest thought to be the burial places of saints, though their names are not generally known.

Another moral fable relates to the Bhilawa or Marking Nut tree (Semecarpus anacardium). Van Gujjars (and it is said, buffalo) avoid this tree, its fruit contains an oil which is blisters sensitive skin. The Van Gujjars are aware that only certain individuals are affected, but they interpret this as

147

differentiating good from evil. Thus, those of pure heart are said to go unharmed, whereas the unrighteous are afflicted by boils even if only the tree's shadow falls across them. Evil Eye beliefs are widely accepted, as they are throughout Northern India and beyond. They do not appear however, to play a major part in sanctioning behaviour, though protective talismans are frequently worn. In principle, the Evil Eye could account for such occurrences as the sudden drying up of a buffalo's milk. When that happens a Holy Man may be called, but at the same time traditional or allopathic medicines are used.

Spirituality and the Van Gujjar World View

The Van Gujjars' world view is founded on their forest-dwelling lifestyle, and has evolved as a unique combination of Islam and Hindu influences. They view humans and nature as partners in the interdependent web of life. They have been influenced by Sufi nature mysticism which merges the glory of nature with the glory of God, and the Van Gujjars commonly use the same Urdu word "Kudrat" interchangeably for both Nature and God. Thus when the Urdu poets say; "Kudrat ka kamaal hai" they are representing both "the marvels of Nature" and "the greatness of God". Because of such usage Van Gujjars can say; "Nature is our God". When Van Gujjars honour the forest and its creatures, they are not only implicitly but actually honouring God. Thus, the rewards of Heaven and the fear of Hell sanction the preservation of forest life, and destruction of the environment is desecrating of God.

Another concept, crucial in reconciling religion with traditional respect for the environment, is that of "Qurbani" or sacrifice. The Van Gujjars use the Urdu/Arabic term, but the Hindi equivalent would be "Balidan" (strength or courage donation) distinct from "Daan", a simple gift, as in a puja donation. Just as Abraham was willing to sacrifice his son, so qurbani represents what one person will do for another, or for Nature, or for an animal......out of love, even to the giving of one's life. Thus, one's generosity to another mirrors the sacrifice of Abraham in microcosm. Similarly a tree is said to "sacrifice" its leaves for the buffalo, and the buffalo its milk. Such sacrifices, whether made by humans or nature, are expressions of love and are natural to the glory of God. When carried out by a person they may be with the expectation of reward both in this life and the next.

The Panchayat and Customary Law

The public meeting of elders, or Panchayat, is still an effective form of social control among the Van Gujjars, and they go to considerable lengths to avoid resorting to outside authority in the settlement of disputes. Membership of such Panchayats is variable, at the lowest level consisting of a few interested elders, but if they cannot settle the dispute or if the matter is of widespread concern, wider membership may be called. At these major Panchayats, which may involve fifty or more elders, only the most authoritative usually speak. In all cases, matters are resolved through discussion, negotiation and consensus. If called before a Panchayat a Van Gujjar has no choice but to attend, and the discipline can be harsh.

Minor matters, preferably settled through private negotiation, include settling damages for injuries incurred in fights between buffaloes from neighbouring deras, forest rights, compensation to be paid as a result of divorce or marital disagreement and so on. Matters of more serious social concern, calling for disciplinary action before a general Panchayat, include; theft, deceit, inappropriate sexual advances, assault, disrespect to elders, damage to sacred sites, desecration of the forest, or any form of violence against God. In recent years there has been difficulty when Forest Department intervention creates disputes, putting the crucial agent outside the jurisdiction of the Panchayat.

Typical judgements meted out by Panchayats range from fines of butter, and the arranging of reciprocal marriage bonds between feuding families, to public humiliation by eating from a dog's dish for a specified number of days, beatings, stoning and banishment from the community.

Poaching and the Desecration of the Forest

When asked about the consequences of poaching or of desecrating nature, most Van Gujjars are certain they would be hauled before the Panchayat, and punished appropriately according to the law of equal retribution. Thus, the punishment for the wanton killing of a wild animal should be death, or at least social death in the form of ostracism. At the same time, most are incredulous of such desecration, and deny that such an abomination could ever happen. And, indeed, we have been unable to find any such instance in spite of extensive enquiries. Accusations of poaching made by forestry officials are so obviously in their own interest, that it is difficult to take them seriously without actual proof. Poaching occurs widely but Van Gujjars contend that they can prove that it and illegal tree felling are done by others, and that if given authority they could work to put an end to it.

We did hear one story about a Van Gujjar causing the death of a wild animal which came to us in various forms, and which could not be confirmed, about a young man called 'Allo' who suffered ostracism. Today he is said to live in an area remote from other Van Gujjars. It appears that in his youth this man had once thrown a stone at a peacock, killing it, So terrified was he of the consequences that he ran away and hid in the jungle for many days. Later a leopard killed two of his goats and in revenge he either killed, or had villagers kill, the leopard. For this he was beaten and banished from the community.

It is sometimes argued that if Van Gujjars kill and skin goats then why not deer or tiger? This however, is an outsider's point of view, as for the Van Gujjar there is a categorical distinction between goats and forest creatures, previously explained. Thus, their behaviour towards goats cannot be extended to other animals.

The Maintenance of Traditional Values

The Van Gujjars are still essentially people of the forest, thankful for their distance from the evils of village life. Yet in recent years their lives have been made difficult by blackmail, bribery and the corruption of those outside, both in the hills and forest. Common land is increasingly encroached upon, and everywhere their grazing rights are being questioned (Gooch, 1992 & 1995). Migration, for many, has become an arduous journey. It is no wonder that many dream of a life free from harassment, whether it be in the forest or settled somewhere with ample land. At this extreme, one disenchanted old man confessed that once his mind had been clouded by the enchantment of the forest, but the forest had brought him only grief and bitterness. Van Gujjars recognise that there are not enough resources for all members of the coming generations in the forest, yet they also define their identity in relation to their forest home and cannot actually visualise life without it. Though some speak of wanting agricultural land, upon questioning they land next to the forest, where they can keep their buffaloes free from harassment. Despite their daily interaction with outside society and manipulation by forest officials, this top priority - to respect each other remains with the honour that comes through the traditional forest lifestyle. "We belong to the forest just as the wild animals belong here. If you remove us, we will be lost. We are born here, and we wish to die here."

Conclusions: Ecological Values and Community Management

The Van Gujjars seem to be peculiarly well fitted by their culture and past history to assume guardianship of the forest. Practical, realistic and down-to-earth, they possess an intimate knowledge of wildlife and the forest. Cultural values and a feeling for the forest are still largely intact, and these speak strongly against poaching or destruction of habitat. They are enforced by sanctions acting in concert, at levels ranging from social pressure and the need for self-respect, to the law of custom enforced by an effective system of Panchayats, also to religion and supernatural sanctions.

Van Gujjars are motivated largely by the need to obtain and keep the respect of their fellow Van Gujjars. A prospering forest is an important avenue to such respect, whereas its destruction or the killing of forest animals is a sure road to dishonour and even community banishment. Influenced by a form of Islam which merges concepts of nature and divinity, God and the natural world are seen as one. Thus respect for nature becomes respect for God, with the expectation of reward, not only in this life but in the world to come.

Such values are resistant to change and presumably would be retained into the foreseeable future, provided that these people are permitted to continue as a close-knit society. Ultimately, some may wish to leave the forest if they are given attractive opportunities which seem viable in relation to their cultural identity. Those who remain will be self-selected to maintain the traditional values and attitudes towards the forest which favour conservation and the protection of forest resources.

Chapter 7
Van Gujjar Problems, Realities & Perspectives

Although the Van Gujjar sense of honour, traditional lifestyle, and community are most important to their identity, they are a direct, pragmatic and realistic people who see the increasing pressures on their culture and lifestyle. They remain extremely confident and proud within their own community, yet they are frustrated by their powerlessness and their exploitation by outside forces. They refuse to give in or shed their values in order to take another path, but they see little hope of success within the existing system as exploitive forces line up against them. These realities have provoked much discussion in recent years spurred on by the escalating park and people conflicts.

Throughout our extensive field interviews, three key themes continually emerged which Van Gujjars see as major threats to the development and survival of their community: (1) a fundamental insecurity and exploitation with respect to earning their livelihood, (2) poor education, and (3) poor health care. Moreover, they are not prepared to give up their identity in return for the solutions offered by others, hence their continued refusal to move out of the forest. They want a secure livelihood, adequate education, and good health within their own framework and culture. This chapter will explore their view of the threats, realities, and potential solutions with respect to each area.

Security, Land Tenure, and Livelihood

The foremost concern of the Van Gujjar people is the profound sense of insecurity they now feel about their livelihood. They have no secure land base, and their pastoral economy is dependent on arbitrary and powerful forces over which they have little control.

According to Dr. Chatrapati Singh, Director, Centre for Environmental Law, Delhi "Historically, forest dwellers have not owned forest in the modern legal sense, but have had occupancy rights. However, these rights were recognised by the British Government under the Forest Act of 1878. After independence the Indian Government did not make any attempt to grant rights to traditional users, and retained State Control over forests. This arrangement contradicts article 39 (B) & (C) of the Indian Constitution which seeks suitable distribution of land and resources."

Land Insecurity and the History of Conflict with the Forest Department

The Van Gujjars have been subject to the loss of customary rights and increasing restrictions in their movement over the past century much as other indigenous groups across India. The state has exerted control over the forests, and caused the Van Gujjars to depend on government regulation in order to maintain their livelihood. Their sense of ownership for the forest in a legal and practical sense has disappeared in the face of Forest Department authority. They retain however, a strong emotional bond with their forest home.

With the notification of the proposed Rajaji National Park in 1983, the policies of the park authorities since that time have brought insecurity that had been slowly eroding up until that time. With the initiation of the park, authorities rapidly put together a proposal for the forced removal and rehabilitation of the Van Gujjars to a new colony to be built in Pathri, a remote strip of eucalyptus plantation in the midst of the Northern plains several hours from the forest.

This plan was implemented and the colony built without consulting the Van Gujjars. It involved the expenditure of 300 lakhs on 512 small cement dwelling units and nearby cattle sheds consisting of a feeding trough and a roof. Each of the 512 licensed Van Gujjar families were also to receive 1.5 bighas (less than half an acre) of land on which to grow food and fodder. There were to be hand pumps, electricity, a school, and tube wells for irrigation. But the facilities were flawed and it seems unlikely that they could have cost 300 lakhs, although the money has been spent in one way or another. Van Gujjars who visited the site reported that the concrete dwellings cracked soon after they were built, and the cattle sheds had cardboard roofs that rapidly disintegrated. The electricity was never hooked up nor the irrigation facilities completed. The decision to allocate 512 parcels of land was based on the outdated 1930s census of Van Gujjar families. Moreover, the amount of land allotted would not come close to meeting the needs of a family for food, much less the fodder needs of buffalo. There was no access to fodder unless they were to buy it, which was impossible given their financial situation. The hot plains climate was wholly inappropriate for a breed of buffalo used to high mountain pastures in summer.

Needless to say the Van Gujjars universally condemned the settlement, and refused to move despite threats and incentives from the Forest Department. At one point a number of Van Gujjars were taken there on a tour by the Forest Department and told to place their thumbprints, as they were illiterate, on a paper to show their attendance. The paper actually indicated their acceptance of the colony, though the agreement could never have been enforced given the Van Gujjars' hostility to the re-location. One man asked "why should we go to prison when we have committed no offence?" (Gooch, 1994, p. 5).

During the 1980s there was a series of legal cases in which the Forest Department tried to force the courts to relocate the Van Gujjars in the Pathri Colony despite their rejection of the plan. Although the court initially supported the possibility of forced rehabilitation, it condemned the plan as inadequate and issued stays against removal of the Van Gujjars. The status quo has been maintained since then with the additional agreement of the Government of Uttar Pradesh that it would not support forced removal of the Van Gujjars.

Throughout the 1980s and into the 1990s the park authorities used threats of eviction as one means of harassing the Van Gujjars. The strengthening of the Wildlife Act through an amendment in 1991 resulted in a ban on grazing in national parks, and served as a pretext for the park authorities to make good their long term threat to evict the Van Gujjars. By force in September of 1992, upon their return from the hills, the authorities denied them re-entry to the park area and their winter homes. This action provoked widespread anger and a period of deliberation among the Van Gujjars as they considered their options over approximately ten days. Finally as buffalo started to weaken from lack of fodder, and the media was drawn into the conflict, the Van Gujjars stated that they would block the major Dehra Dun-Delhi Highway if they were not allowed to re-enter the proposed park area. Facing the proposed blockade and media condemnation, the authorities relented on a temporary basis. This provisional re-admittance was extended the next year, and park authorities now indicate that they will not forcibly remove the Van Gujjars from the area, although the park plan continues to state that their removal is essential. No strategy for their removal is suggested.

The Forest Department has recently made new attempts to persuade the Van Gujjars to move to Pathri. They have been "successful" to the extent that approximately 65 "families" began occupying space there in the summer of 1995. They report that only one of those families however, is complete. The remanded are individuals who do not have official forest permits and whose families have stayed in the forest. The one family which moved is reported to have done so due to intensive harassment from a wealthy Van Gujjar, who was receiving favours from the Forest Department.

The main incentive for the group to move was an informal agreement from the Forest Department that the land allotted for 512 families could be divided temporarily among those present such that they each received roughly 10 as opposed to 1.5 bighas of land. Although some buffalo initially went to Pathri, a number of them died and the rest have returned to the forest. There were informal agreements between the Van Gujjars and the Forest Department to allow the buffalo to remain in the forest. The Van Gujjars were also promised that the electricity would be hooked up, the irrigation facilities completed, the school opened, and titles to larger pieces of land given to them. None of these promises has been fulfilled in fact, as of February 1996; the Van Gujjars are unhappy and threatening to leave according to a spokesperson we recently interviewed. Van Gujjars we interviewed in the forest view those who moved with pity and predicted the failure of the move. They expect that the group will soon return to the forest. This latest episode simply reaffirms to all Van Gujjars the failure of the park authorities to negotiate in good faith. They view it as a desperate attempt to demonstrate that Rs. 300,00,000 had not been wasted, and also a means to fend off other interests who might have another use for land that has been unoccupied for nearly a decade.

Based on the events of the last fifteen years, it is not surprising that the Van Gujjars have a severe sense of insecurity about their land base, and a profound hostility and distrust towards the park authorities. It is a testament to their positive social values that there has not been physical conflict yet.

Land Insecurity and Forest Management

This insecurity in terms of whether or not they are to be evicted, is increased by a range of problems which threaten Van Gujjars on a day to day basis. There is the current possibility of losing a portion of their forest or being shifted to another forest. For example, one family reported that they were moved out of a khol a number of years ago as it was being closed for several years. When it re-opened, they found that their forest permit was diminished by one-half, and the remainder had been given to another Van Gujjar who was friendly with forest officials. Van Gujjars report that adjustments in the forest are made in the summer in favour of a select few, while nomadic Van Gujjars are in the mountains.

A number of different deras reported that a couple of Van Gujjars under the influence of the forest officials played key roles in pressurizing those who recently moved to Pathri. In return, they have received access to forest land in a closed khol. We also met a local bhabbar grass contractor who reported that he was an advocate for the Van Gujjars in that khol with the local forest officials. The Van Gujjar family reported however, that the contractor periodically demanded a fee for his services, and if he was not paid would use his influence to have false charges brought against the dera.

Survival in this management structure results in some Van Gujjars playing the system. Thus we met one Van Gujjar who was lodging a complaint with the Forest Department about another one who was living in a closed khol. Van Gujjars who are dependent on Forest Department favours to support large herds are locked into continuing the practice in order to maintain their lifestyle.

Land Insecurity and Problems in the Hills

Another element of Van Gujjar insecurity over land is difficulties in their summer pastures and on the migration. Despite management policies in the proposed park area being against migration many Van Gujjars still have permits in the hills and maintain their annual trek. The research team made two trips to the hills and several visits along the migratory routes in order to identify problems. It should be noted that land difficulties noted below, although important, are only one aspect of problems in the hills. Many villagers welcome Van Gujjars in the summer only to the positive mutual trading positive mutual trading relationship.

Issues in the hill areas seem to vary with locale. Van Gujjars who go to Himachal Pradesh report harmonious relationships with the local people, who value their presence as a source of milk. They report sufficient but limited pasture due to the presence of new apple orchards. The forest officials do not usually interfere with their lifestyle.

Van Gujjars staying in the Harkidun region reported some conflicts with villagers over use of pastures, and supposed conflicts between Van Gujjar and local animals, which Van Gujjars viewed as simply an opportunity for local mafia to extract money from them. They indicated that the total amount of land was not much an issue, as the inability to resolve conflicts over parts of it. In particular, the officials who live and work in the area are seen as always siding with the villagers, and thus Van Gujjars have no impartial authority to assist in the resolution of disputes. Moreover, locals can threaten to use the justice system on which they also have influence. In addition, there is yearly encroachment by landlords who are turning forest land into apple orchards and crop, in return for fees paid to revenue officials to change the official land status.

Further to the east in the Uttarkashi and Chamoli districts, Van Gujjars report similar problems. Several Van Gujjars reported that wealthy rural landowners, not the local villagers, were the source of trouble. These landholders contract shepherds to look after large cash generating herds in the summer pastures. These landowners have particular influence with local officials and authorities which makes it more difficult for Van Gujjars to receive justice. In this area, Van Gujjars report there are few direct difficulties with forest officials except that they do not venture into the higher areas to investigate problems. They must also pay fees to protect their deras from being destroyed in the winter.

Despite the problems and conflicts, Van Gujjars report less difficulty with the forest officials in the hills than in the park area. There are also positive relationships with villagers based on economic ties, Van Gujjars sell milk and milk products while buying food and supplies. The land insecurity involves worry as to how much pasture there will be and what conflicts may result, rather than whether they will be evicted from the forest. Moreover, the hills have advantages in that the fodder is generally superior and the price of milk is higher. The Van Gujjars also benefit the forest in the hills as they discourage illegal tree felling. We observed illegal felling in progress below their lands, but not within their zones of control.

The major source of insecurity in the nomadic movement is the increasing difficulty in finding forest land to stop on along the path and difficulties in passing through villages. While many villagers welcome the Van Gujjar's presence for the economic benefits, and for the manure the buffalo can spread on open fields, large landowners are still encroaching on forest lands, and either preventing Van Gujjars from using the forest or charging them a fee for doing so. We travelled with one set of Van Gujjars who were required to pay Rs. 1000, the regular yearly village taxes, in order to pass their herd through the village. Other Van Gujjars report sneaking through occupied lands in the dark for fear of being caught, and locals demanding payment in butter or ghee. Van Gujjars also noted however, that due to their increasing literacy in the past couple of years, they now have more confidence in

159

dealing with officials and, in fact, laid a complaint in one case. In addition to stopping the problem in the individual case, it sent a message to the locals that Van Gujjars had some legal recourse, which reduced the level of harassment there for that season. They also mention that they have recently received support in a couple of instances from DFOs in the area, thus sending a positive message to the individuals who had been harassing them.

Some Van Gujjars have taken to renting a fodder truck to accompany them during parts of the migration where grass is particularly scarce, yet this means additional cost. Increasing migration costs are most difficult for poor Van Gujjars with small numbers of buffalo, as they produce less milk and generate less cash. If the obstacles increase, it becomes more sensible from an economic point of view to remain in the Rajaji area. Although smaller herds have less impact in Rajaji in the summer, they remain a drain on the ecosystem at a time when regeneration is important.

In short, the major sources of insecurity about migration come from the failure of authorities in the hills to maintain commons forest lands. There will, inevitably, be disputes between Van Gujjars and villagers, and there the problem is a need for impartial arbiter in these disputes.

Economic Realities and Insecurity

Van Gujjars are not suffering from abject poverty and can earn a marginal income through their herding which allows them to maintain their way of life. Their economic insecurity however, results from their powerlessness against middlemen and forest officials who demand arbitrary payments as they see fit. Table 4 describes the distribution of families by forest range with respect to the size of their buffalo herds, a measure of economic status as it directly relates to income level. The most frequent herd size is between 10 and 20 buffaloes, and there is relatively little disparity across the distribution. Two-thirds of families have less than 20 buffalo, and only 13% percent have more than 30 buffalo.

Table 11 provides one Gujjar's financial report for his family of seven people with a permit of 14 buffalo, an average herd size. Though it relates to one family, the data was cross-checked by asking similar questions of two other families to validate the findings. The data indicate that during the winter season, with 14 buffalo, the family would be in debt by approximately Rs. 5000 accounting only for food and herding expenses. 30% of the family income goes on fees and taxes to forest officials on a regular basis, roughly the amount by which the family is in debt. In addition, officials require milk or butter for special occasions such as family weddings or visits from important higher officials. We interviewed a man responsible for collecting the milk payments from deras on a daily basis in another area. He said he collected approximately 20 kg of milk per day to be divided among 15 or 16 officials including himself. He stated that he collected between 5 and 10 kg of butter or ghee for the visits of dignitaries depending on the person's status.

The high winter debt reported by this family is not unusual, and most families are dependent on and exploited by the milk middlemen to obtain loans for this season. In this instance the Van Gujjar did have buffaloes in excess of the permitted number although we were not told the figure, hence his actual income was higher than reported (he indicated that there was an additional fee of Rs. 60-80 for buffaloes beyond the permitted number along with an additional one kg of butter per season). In addition, the economic returns in the summer season are somewhat higher and thus provide an opportunity to recoup some of the winter losses. Milk prices are higher in the summer both in the hills and the Rajaji area. Financial burdens are less in the mountains as well as in the park area. This Van Gujjar reported that fees in the park area were 50% lower in the summer season. On the other hand, milk production is reduced in the summer in the park area. Nomadic Van Gujjars have higher expenses due to the migration, but milk prices are higher in the mountains, milk production is good, and there is no need to supplement the buffaloes' diet with oil cakes and bran as pastures are good. There is a greater advantage in maintaining a nomadic lifestyle for those with larger herds, and poor families with very few buffalo have more incentive to remain in the park area in the summer.

Table 11
Income and Expenses for a Van Gujjar Family with 14 Buffalo

Income

12 kg	of milk per day for sale (2 kg used by family)*
Rs. 7.0	per kg (range is from 5.5. to 8.5 depending on the distance and marketing system)
Rs. 84	is daily income for 180 days in winter season
Rs. 15,120	rupees of income per season

Expenses

Rs. 4200	Oilcakes for 6 months: 12 sacks, 25 kgs per sack, Rs. 350 per sack
Rs. 2400	Bran for 6 months: 12 sacks, 50 kgs per sack, Rs. 200 per sack
Rs. 3000	Flour for 6 months: 100 kgs per month, Rs. 5 per kg
Rs. 900	Sugar for 6 months: 10 kgs per month, Rs. 15 per kg
Rs. 960	Tea for 6 months: 2 kgs per month, Rs. 80 per kg
Rs. 1584	Dal for 6 months: 12 kgs per month, Rs. 22 per kg
Rs. 1200	Vegetables, Salt, Masala, Chillies, Rs. 200 per onth
Rs. 1200	Clothing and other, Rs. 200 per month
Rs. 4688	Taxes and Fees to forest officials (see below)
Rs. 20,132	Total

Fees to Forest Officials

Butter

6 kg	per season to forest guard, 1 kg per month
3 kg	per season to forester, 1/2 kg per month
<u>14 kg</u>	per season to the ranger, 1 kg per buffalo per season
23 kg	Total winter fee for butter, value of Rs. 80 per kg
Rs. 1840	Total value of butter per season

Milk

1.75 kg per day for 14 buffalo, valued at Rs. 7.0 per kg
Rs. 12.25 per day value for 180 days
Rs. 2205 total value of milk per season

Grazing and Lopping Tax

Rs. 35 per buffalo for 14 buffalo
Rs. 490 for season

Rs. 1840	Value of butter fee
Rs. 2205	Value of milk fee
<u>Rs. 490</u>	Grazing and lopping taxes
Rs. 4535	Total fees and taxes to forest officials for season
Rs. 15,120	Total income for season (from above)
30% of Income goes to fees and taxes	

* Buffaloes produce more than 1 kg of milk per day when milking but this average figure has proven reliable over time in the self-help coop milk program as a prediction for supply. It accounts for the range of factors such as non-milking buffalo and presence of calves and bulls in the herd.

The financial picture for this family shows their attitude towards forest officials, and their need to exceed their permitted level of buffalo. They are contributing nearly a third of their income to park officials, which roughly equates with their winter season losses. Maintaining a herd above the permitted number is essential to their survival. The cost of these fees is born by the forest through increased herd size. At the same time, they see forest officials collecting fees from a wide range of other people who are extracting resources from the forest (poachers, tree fellers, grass cutters, etc.). Moreover, they are always subject to the arbitrary judgements of forest officials who can extract fines for not following approved forest practices. One Van Gujjar reported having to pay Rs. 500 extra for the tractor delivering oilcakes to his dera to enter the park area. The fact that they may have more buffalo than permitted, and do not lop according to the official definitions, provides ample flexibility to extract more money. Yet these actions are necessary in terms of surviving under the present system.

The specific form and level of forest payments varies from area to area, but the typical range was between Rs. 300 and Rs. 400 per year per buffalo, in either cash or milk products. One woman reported that seventeen deras paid Rs. 15,000 for the season in addition to providing daily milk. In this area officials were demanding payment in cash rather than milk products, which is a greater hardship for Van Gujjars. Fees go up in instances where there are additional restrictions, such as a ban on lopping.

Some use these figures to argue that the Van Gujjar lifestyle is economically unsustainable in the long term, particularly if the forest continues to deteriorate and milk production is reduced. This is the root of Van Gujjars' present sense of economic insecurity. They recognise the bleak scenario under the present system but they do not want to abandon their valued lifestyle and culture despite of some who view them as "backward" and in need of "development". There must be a change- either they are forced to abandon the forest and their lifestyle in the long term, or the present system must change to eliminate the massive outlay of money for fees. There must also be steps to improve and protect the forest. They argue that blaming the Van Gujjars and demanding their removal is simply a means to shift blame from the actual problem, an enforcement management system which has been antagonistic to local people.

Insecurity and Relationships with Milk Middlemen

Traditionally Van Gujjars have relied on milk middlemen in order to market their milk in towns and cities. The middlemen hire people to collect and transport the milk on a daily basis and sell it through commercial outlets. Part of the arrangement is usually that the Van Gujjars will buy flour and oilcakes from the middlemen at inflated rates. In turn, they can freely obtain loans which become necessary given their economic status and lifestyle. The result is that they are at the mercy of these middlemen who give them a lower price for milk than they could receive in the local market. In this way they are paying interest on their loans in an informal manner that they cannot easily quantify. There is always the threat that the middleman could ask for repayment of the loan, so there are few options for individuals to leave the system. Recently, Van Gujjars have organised a co-operative milk marketing scheme through RLEK which is increasingly successful, offering a significantly higher price for milk (Rs. 8.50 per kg) than they have ever received in the past. It is still difficult however, for some Van Gujjars to join the program as they are beholden to the middlemen for the loans.

Directions for the Future

The bleak realities outlined above have caused much discussion among Van Gujjars as to future directions which have resulted in two of approaches- either to request land, or community forest management. Until recently, the request for land has been favoured because there is a history of land offers so it seems more feasible. Both responses arise out of the fundamental problem of land and livelihood insecurity, and amount to flip sides of the same coin. Van Gujjars recognise that they must control their land base and livelihood.

(a) Security Through Agricultural Land. The land option was initiated in the 1970s with various government offers and recommendations based on this approach. A group of Van Gujjars was settled in Dhaulatappar near the Yamuna River, and a number of other Van Gujjars from Ranipur and Motichur received land as a result of the arrangement to give up their permits in the hills in the 1970s. The Bhasin Report in 1979 commissioned by the Uttar Pradesh Government laid out a proposal to settle Van Gujjars along the borders of the park.

These efforts have generally failed due to lack of land suitable for the Van Gujjars, their failure to abandon their traditional lifestyle even when they receive land, and unfulfilled promises by the Forest Department in negotiations. The Bhasin Report was never implemented because the land allotted was not, in fact, available and the cost was enormous, though it did meet the Van Gujjars' requirement of land adjacent to the forest. The Van Gujjars who received land in the 1970s have nearly all returned to the forest, when they were forced to choose between land and forest access in the park area. Those who remained have never received papers giving them ownership of the land they are occupying, thus their tenure remains insecure. Some of the Van Gujjars who have settled on land continue to be nomadic, including several wealthy Himachal Van Gujjars who own large apple orchards and yet maintain a nomadic lifestyle while others run the orchards. Pathri was rejected outright for all the reasons previously stated. The failure of the Forest Department to keep promises to the small group which recently went to Pathri has reconfirmed that land promises cannot be trusted.

Despite the failures, one often hears land as the initial request when a Van Gujjar talks with an outsider who, they believe, may have some authority or influence. Thus, one Van Gujjar woman was adamant in her request for land at a recent meeting although she had previously received land, but was not using it and had returned to the forest as a nomad. Our careful questioning clarified that they would only be satisfied with adequate land on the edge of the forest where they could grow a few crops and continue to use the forest for the buffalo. In a number of instances individuals suggested that the best land would be right outside the khol in which they presently reside. They realised however, that such unoccupied land does not exist, and that Van Gujjars have generally not remained on land when it has been allotted. Their requests for land are also reinforced by the notion that accepting land need not involve losing the forest as a resource for their buffalo. Under the current system, they can always pay fees to forest official and continue to keep the buffalo in the forest.

The requests for land seem most consistent and heartfelt from people living in the Ranipur Range, Motichur, and some parts of Kansrao. Generally these are among the poorer Van Gujjar families who do not migrate to the hills in the summer. Their forests are poor and this section of the proposed park area has the highest level of pressure from villagers, tree fellers and poachers. Despite their requests, however, they see no viable option for land, given the history of the issue with the Forest Department.

(b). Security Through Community Forest Management: A second approach which has grown in strength and favour in recent years is the request by the Van Gujjars to manage the forests.

It is a large leap from facing eviction to asking for management of the forest. Van Gujjars however, have never doubted their ability to manage the forest, but it is only recently that there has been recognition in official circles that the existing approaches are failing and community alternatives are needed. The Van Gujjars see community management as a viable alternative for addressing their insecurity, in that it would provide them with control over the land that they use and a very different and more productive framework for their relationship with the Forest Department. They recognise that the forest is deteriorating under the present system, and that their cultural survival and the forest go hand in hand.

(c). Security through Milk Marketing: The self-help milk marketing scheme is seen as a very viable means of providing leverage in the marketplace to relieve their economic insecurity, regardless of whether an individual gives milk to

the program or continues to utilise milk middlemen. Competition from the milk marketing scheme which is presently collecting 26 quintals of milk per day and selling it at Rs. 8.50 per kg has resulted in middlemen raising the price they offer for milk for fear that Van Gujjars will desert them. Many Van Gujjars are restricted from giving milk to the co-op because of their loans to the middlemen, but some are now dividing their milk between the two systems. The combination of increased security, through the co-op scheme and not having to pay fees under a community forest management plan, offers a promising direction for them.

Education
Problems and Progress

The second major issue for Van Gujjars is their lack of education, as they recognise that education is an essential means through which they can gain power in their dealings with the society. Until the initiation of the Van Gujjar literacy program through RLEK in 1992, nearly all Van Gujjars were illiterate, giving them a severe handicap when dealing with villagers, middlemen, and local authorities. The program prepared primers, based on Van Gujjar words and themes, which became a source of pride as they learned to read. As many as 350 volunteers lived with Van Gujjars in the forest, and taught informal classes in deras in accordance with the time schedules of their students. There was tremendous commitment to the program; one Van Gujjar elder collected small sticks for kindling during the day to provide additional firelight in the evening for the family classes. In turn, the volunteers, often from local villages, gained a new perspective on the Van Gujjars and life in the forest.

Avdhash Kaushal, Chairperson, RLEK feels that "Education programmes for nomadic people around the world set forth that settlement is a pre-condition to their being made literate. We at RLEK, however, do not agree with this and believe that any programme for them must necessarily include their traditional ways and traditional wisdom. It is only the tired and the weary that settle down and the fact that this community has not done so over thousands of years goes a long way to prove the endurance of their life style."

More than 15,000 Van Gujjars in the wider region have completed the three primers over the three- year period, and have achieved a basic level of literacy. One of the major problems with the program involved disputes over who would get volunteer teachers and where they would teach as there were not enough to go round. Another difficulty has been the short term nature of the original program such that those who have achieved a basic level can easily slip back. This literacy process has heightened awareness of the value of education, and increased desire to proceed beyond the basic level. Van Gujjars now want to ensure that their children are given the opportunity for education which they themselves did not have.

On one hand Van Gujjars see education as a means of offering alternatives to their children who may not wish to stay in the forest in the future. A second reason for education however, is to equip themselves with a tool for fighting for their rights and their lifestyle in the forest. Although they have been very enthusiastic about literacy teachers in their deras, and in supporting development of their first school in Dhaulatappar, they have not been eager to send their children to village schools that have been accessible. The concept of a self-contained community that teaches its children values and culture remains a dominant theme of their discussions. The group that recently moved to Pathri is sending only a few of its children to the village school although increased access to education was stated as primary reason for their move. They continue to press for their own school in the colony.

Directions for Education

Education is a priority as it provides a vehicle which will allow them to interact on equal terms with the outside world, make choices, and defend their lifestyle. They note however, that they pay a large percentage of their income to the government, and yet do not receive any services in return. They want access to educational opportunities that reinforce their culture and perspectives. A Van Gujjar from the Mohand

Range noted, "My children need education as some may move to the city to obtain jobs while others remain in the forest with our traditional occupation."

Health

Problems and Progress

Good health and accessible health care, both for people and animals, is the final priority area which is commonly voiced by Van Gujjars. Van Gujjars see health as connected to their lifestyle in the forest.

Mustooq Lambardar, a Van Gujjar, says that "children and livestock are much healthier here in the forests than in the cities. The atmosphere in the cities is polluted. People who live in the forests get pure air from the sky. Water is also pure... We people do not eat meat, drink wine, or gamble. We do not rob others and steal... We depend on the blessing of God and we get everything...The children of Van Gujjars are under control of their parents and move on the honest path only."

They feel they are in better health than the people in the rural villages and towns whom they meet. They note that they do not suffer so frequently from major diseases such as tuberculosis, cancer and leprosy. They also define health as going beyond the physical aspect to include one's attitude and behaviour in life.

They have some traditional remedies for illnesses, but as a pragmatic people they are open to whatever approaches they feel are effective, and therefore, have been quick to embrace allopathic approaches as compared to other traditional people. They also recognise that, due to their remote locales in the forest, they have had less access to health care than others. Some families in khols where water is scarce are particularly concerned about the quantity and quality of their water supply. The childhood mortality rate has been high due to common diseases; hence there is much support for vaccinations. An increasing awareness of the limits of the forest to support people, and the recognition of the health drain and stress on women who have large numbers of children, has also brought about requests for family planning.

Finally, there is the request for veterinary support and vaccinations for their animals as their buffalo are the only means of survival in the forest.

Until recently, Van Gujjars have been torn between viewing the forest and their lifestyle as bringing good health, and recognising that their remote homes made access to modern health care difficult. In the last year, however, RLEK has begun a health and veterinary program whereby a health team visit selected sites in the forest on a regular rotating basis. The health program has also begun a training process for Van Gujjar paramedics to provide basic services. In the past three months, the health team has initiated a broad childhood vaccination and family planning program, which has received high participation levels among Van Gujjars.

Directions for Health

The success of the health program has demonstrated that Van Gujjars can maintain their health through their forest lifestyle, while still availing themselves of modern health care when it is required. They would like to see the health program increased in scope and coverage, particularly when they are in more remote hill in the summer. They would also like to have an improved means of communication so that they can call the health team to a medical emergency. They support the overall approach, and see the need for developing it further.

Synthesis of Van Gujjar Needs

Van Gujjars recognise the widespread difficulties that they face under the present forest management system. They view their culture and lifestyle as their greatest strength and refuse to accept the notion that it is backward, impoverished, and should be abandoned. Their priority is to maintain this lifestyle which means addressing their demands for security in land and livelihood, access to education, and good health care. They want to control programs and have them delivered in a form that support of their own culture and values. They have seen effective models for education and health programs in the forest.

Their ideal is to own land on the edge of the forest, grow some food, and use the forest for their buffalo while continuing to migrate to the hills in the summer. They are also realistic however, in admitting that such land is not available. They have always believed that they would be far better managers of the forest than the Forest Department, and cite endless examples to put their case. It is only recently however, that they have begun to see that this is feasible given the continuing problems with the present system and the support they have received from outsiders. Community forest management has become the preferred option as they see that it could maintain their lifestyle, change the management system that has put it in jeopardy, and preserve the forest.

Chapter 8
Background on Taungyas and Local Villagers

Taungyas

History of the Taungya System

By the early 1930s the Forest Department recognised that its existing system of growing and managing forest plantations was unsuccessful, and was looking for alternatives. Labours were invited to the area in the early 1930s under the sponsorship of the Forest Department. The arrangement involved the Labours planting and maintaining plantations, and growing crops for themselves for the first few years in between the rows. As the plantations grew up, they would move their crops and start a new plantation. In addition, they were given a small amount of land for traditional agricultural purposes (1 acre per family), and rights to grazing and the extraction of fuelwood and bhabbar grass from the forest. This system of plantation was called the Taungyas System and the labours involved as Taungya people. Since the Forest Department however, viewed them as a transitory work force, they did not have the right to pucca houses and their villages were not accorded revenue status, thus barring them from access to development funds, local voting rights, and a gram sabha.

The Taungyas were no longer needed by the Forest Department after the end of commercial forestry operations in the area and the initiation of the proposed park in the 1980s. The Forest Department denied that they had any permanent settlement rights, and demanded that the four villages within the proposed park boundary were vacated. One village had been forcibly evicted to an area outside the proposed park, before the remaining villages could file a case in the Supreme Court of India to block further evictions. The court did, indeed, stay any further evictions, but also rejected the Taungya plea that they be accorded traditional land rights and revenue village status. The court indicated that this decision was in the hands of the Forest Department which owned the land. Thus, the Taungyas remain in limbo, secure from eviction, but unable to benefit from development programs, and exercise full legal and political rights.

Taungya Perspectives

Like the Van Gujjars, the Taungyas spoke of a deep antagonism toward the park authorities during our meetings with them. They argued that they had given up their rights and property in their original villages to come and work for the benefit of the Forest Department. They had provided fifty years of service, and now found themselves without ownership of the land they had nurtured, and without legal rights. After fighting off eviction, they still have lost their rights to bhabbar grass, and continue to extract it only by paying regular fees to park officials (10 rupees a headload). They indicated that each village had a limited number of khols from which they obtained bhabbar grass based on proximity and traditional practice. They also complained that wild animals, particularly elephants, are a constant threat to crops and sometimes humans. They view the animal problems as having increased since the initiation of the proposed park. The elephant population has increased, but their traditional migration routes have been blocked by government encroachments in the elephant corridor. There is no effective means through which the Taungy can receive compensation for their crop losses. They also repeated the Van Gujjar view that tree felling and poaching occur systematically, and that their past attempts to report smuggling or poaching had been ignored, or resulted in retribution.

Directions and Needs

Firstly, Taungyas demand revenue village status, legal rights, and access to development programs. They point out that their traditional occupation is to plant and grow trees, and that they have the expertise and interest in growing and protecting the forest. But they need to be guaranteed access to the resulting forest products that they require for their subsistence lifestyle. They are interested and enthusiastic in forming Forest Protection Committees which would be responsible for discrete land areas, and from which they would enforce sustainable resource use. They want their traditional rights to the extraction of bhabbar grass recognised, as well as rights to grazing and fuelwood collection.

Villagers on the Park Periphery

Problems and Perspectives

Villagers on the southern park boundary repeated the Taungya antagonism toward the park authorities, and its attempts to extinguish their traditional rights to the forest. Their foremost concern focused on access to bhabbar grass, which is the mainstay of their livelihood as their agricultural land is very limited. Many are landless and wholly dependent on bhabbar and rope making. They recounted the history of the problems in management of the bhabbar resource by the Forest Department (see Chapter 5) along with attempts by the park authorities to prohibit their use of grass in 1990 and 1991. Although they now have access to the grass, they greatly resent the idea that they must pay to extract grass illegally when it is their traditional right. They also want their traditional rights to grazing and firewood collection recognised although they averred that they use only the edge of the park for these activities. In some cases, the villagers were not using the park area for grazing as it was too remote from their homes and the resource was limited. Some were collecting leaves from the forest and bringing them back for animal fodder. They stated that each village had a limited area within the proposed park which they used based on tradition and proximity.

Villagers also complained about the crop predation caused by animals living in the proposed park area, particularly elephants. They stated that in some cases wild animals made it impossible to plant particular areas, and there was no effective way to receive compensation. They viewed illegal tree felling and poaching as major problems, which park officials were unable to stop. They reported that, roaming the forest were elite gangs with high level political connections, which utilised nearby farms as bases for operations. They admitted that there were a small number of villagers who also engaged in smuggling activities. But, as their previous efforts to report illegal activities had been ignored by officials, they saw little reason to pursue them considering that park officials were part of the problem.

Residents of Rasulpur Village in the Chilla Range had a somewhat different view. Although they resented the Forest Department's policy of cancelling their traditional rights, they were also reasonably supportive of individual local officials who, were doing their best within an exploitive system. They do not have to pay fees for grazing and firewood extraction, although both are technically illegal. They are able to remove grass for thatch and some bhabbar grass in return for fees, but this is not a primary issue here, as rope making is only an auxiliary source of income when grass is available.

The Rasulpur villagers pointed out that the Forest Department had been responsible for the degradation of the forest on the periphery near their village, by clear felling the forest in the 1970s and replacing it with a commercial Teak plantation having little ground cover which was useless to domestic or wild animals. Unlike other villagers, they felt that their reports of illegal activities to forest officials were often followed up, and that in general there was little tree smuggling and poaching in the area. They also felt this was due to the fact that their villages were right on the park boundary and that they had a high level of presense in the forest to a depth of two or three kilometres beyond which were the Van Gujjars. They affirmed that they still had a stake in the forest despite their loss of rights, because they were generally allowed access to forest resources without harassment from local officials. They noted that access however, was dependent on individual officials, and that there were problems when officers occasionally became "irritated".

The Rasulpur residents also complained of crop predation by elephants and that there was no way to receive compensation. They had constructed wooden fences to protect crops but noted as wood rotted over time it was surmountable by large herbivores and elephants. They suggested that stone walls would be far better, but were prevented from collecting the stones due to park policy. One family we interviewed were of Gothiya background, and had been evicted from the proposed park area in the mid-1980s. They were fortunate in that they already owned land on the border, and thus had somewhere to go. They had refused to surrender their forest permit, and still pay lopping taxes and fees to maintain their use of the forest, despite living outside the proposed park.

Our discussions with several Tehri Dam evacuees in the elephant corridor were rather different. They did not express animosity toward the Forest Department, were generally happy with the land they had received in compensation for their losses in Tehri, but they too, recognised that they were occupying the elephant corridor and had respect for the elephants' needs. They rely primarily on agriculture for their livelihood, and seem less dependent on forest resources. Possibly because they have lived in the area for only a few years they are less resistant to the idea of re-location so long as they receive just compensation and land that is as good or better than what they currently have. They particularly want land in the Doon Valley, and mention that there have been discussions of possibilities with the Forest Department.

Villager-Van Gujjar Interactions

Both villagers and Van Gujjars expressed support for the traditional rights of the other group and sympathy for the harsh treatment they all had received from the park officials. Throughout the field research process we asked groups to define their area of forest use. In almost all instances both villagers and Van Gujjars noted there was little overlap in their areas, and where there was, there was informal agreement as to who had the rights to particular resources (see Map 7, Chapter 5). In a number of instances Van Gujjars lop in boundary areas of the park, while villagers use the area for grazing and fuelwood collection. Unless extracting bhabbar grass, villagers rarely venture beyond a few kilometres into the park.

Van Gujjars recognise the villagers' dependence on bhabbar, but they make a distinction between villagers' traditional rights and large scale extraction by contractors.

They feel that contractors have little investment in the resource, and that they use their access to the forest as an opportunity to carry out illicit activities. Villagers share these views. There was also resentment among Van Gujjars in the Dholkhand Range about the high numbers of villagers extracting bhabbar grass daily from the area through the payment system. They felt that the resource was being over-exploited with no controls, while park officials were accumulating income through fee collection from villagers. Villagers stated that on the other hand there were a small number of affluent Van Gujjars in the area who were cutting some of the best grass before villagers could get it, and then selling it to them.

Van Gujjars across the western portion of the proposed Rajaji Park respected villagers traditional grazing and fuelwood rights, but noted that some villagers were responsible for illegal tree felling, bark stripping, and poaching. They are particularly upset by bark stripping, and the resulting death of Padal trees which are a good source of fodder. Van Gujjars in Ranipur felt that the large population nearby in Haridwar resulted in over-exploitation of the forest, even fuelwood. They also felt that fuelwood extraction was for commercial as well as subsistence needs, and that illegal felling was associated with it. Berkmuller (1986) found that pressure on the forest from the nearby urban area was highest in this region.

Directions and Proposals for Change

All the villagers we interviewed demanded that their traditional rights to the forest be recognised, and all supported the idea of organising Forest Protection Committees (FPCs), through which they could manage separate forest areas regularly used on the borders of the park. Rasulpur residents in the Chilla Range had given particular thought to the workings of a committee. They identified the boundaries for their proposed FPC, and noted that each village had a discrete area of usage so that FPC areas could be designated without conflict. They noted that the more distant villages had alternative forests, and did not usually depend on the park area. In instances where this was an issue, they felt that these villages could work through FPCs near them. They proposed to donate a portion of their crops to hiring a local guard, and asked for training in effective committee functioning.

The villagers we interviewed on the southern periphery varied in to their ability to define how FPCs might function, although they all supported the idea. There appeared to be a greater diversity of groups and interests in some of these villages, which would make the organisation of FPCs more difficult. They indicated that they would need time for discussion in order to respond to our questions about exactly how FPCs would function. On the other hand, villages we revisited in the pilot area were precise about how FPCs would operate, who would be members and how they would manage the bhabbar resources. Their ideas are presented in part IV as the proposals for community forest management in the pilot area zones to be managed by villagers.

Synthesis of Villager Needs

Villagers are more diverse in their characteristics, concerns, and needs than Van Gujjars, yet they are united by their demand for restoration of their traditional rights to the forest, and are willing and eager to participate in forest management. The villagers in Rasulpur and the pilot area had given more thought to how they could participate in forest management. Others, with whom we talked, had less sensitivity to the limits on forest resources relative to the Van Gujjars. They were more likely to indicate that simply restoring their traditional rights would solve at the problems, despite the present deterioration of the forest. We believe that this attitude is in part due to these villages being far from the forest, so that they are more likely to conceptualise only in relation to their own needs. It seems however, that there has also been less detailed discussion of alternatives for the future, and thus less consideration of the problems in community management. In some cases outside the pilot area, precise, detailed proposals for community management of forest areas, where villagers are the primary users, will take more research and discussion in the specific locales.

PART - FOUR

COMMUNITY FOREST MANAGEMENT IN PROTECTED AREAS

The national and international policies, the history of forest and park management in the Rajaji area, the characteristics of the Van Gujjars, and their relationships with other local people, indicate that the Rajaji area holds excellent promise as a test site in which to implement Community Forest Management in Protected Areas (CFM-PA). Part IV presents this approach, and begins with an initial chapter defining CFM-PA, including the goals and principles for applying it in specific locales. We believe the CFM-PA approach has the potential to provide an effective management alternative to the present system in a wide range of settings, and, thus, we have provided the rationale and definitions without reference to the Rajaji context in Chapter 9. Implicit in the definition of CFM-PA however, is the need to structure precisely the approach to specific needs of the local communities and the ecological context. Thus, there may be a wide variance in structures from one area to another. It makes sense to evaluate CFM-PA effectiveness initially in conducive settings, and then examine the flexibility in other contexts. Chapter 10 argues that Rajaji is a appropriate test site, and proceeds to define accurate structures through which to implement CFM-PA in this context. Chapter 11 then presents the schedule and local data for implementing these structures in the first year pilot zone within Rajaji.

Chapter 9
Definition of Community Forest Management in Protected Areas

The Need for Change

The problems of the Rajaji exemplify the failures of existing policies for protected areas throughout India (see Chapter 2). These problems are inherent in the structure of a system with limited resources, which has served to alienate the local people whose support it requires. Enforcement activities have not been able to curtail tree felling and poaching, and incentives are such that many parties are encouraged or able to exploit the forest for their own personal gain. There can never be enough restrictions and enough enforcers as long as the system is widely viewed as illegitimate by local people. Local people have lost their traditional right so that nearly any action in the protected area is illegal, and at the same time any action is also possible in return for a fee paid to forest officials. In the process, local people have ceded their traditional sense of responsibility for the forest to the authorities, and respond to the officials' exploitation of it with a mixture of anger and resignation.

It is essential to emphasise that the problems are due to the system of management, rather than to the individuals within it. Given the existing incentives, working conditions, and institutional context, it becomes extremely difficult, if not impossible, to oppose the present system. Structural change would benefit forest officials in carrying out their duties and support the traditional lifestyles of local people. Conservationists, for the several decades past have argued that the government managed parks protection model which excludes local people would be the only means to insure protection of the ecosystem. Across the nation and the world, including Rajaji, that model is bringing about the destruction of the ecosystem it was charged to save. It is destroying the culture and lifestyle of the indigenous people living in and around protected areas. An alternative approach is urgently required.

Limitations of Joint Protected Area Management

Joint Protected Area Management (JPAM) represents an alternative response to the existing model. In this model the Forest Department and local people work as partners in evolving co-operative approaches to achieving protection of the ecosystem and meeting the needs of local people. This partnership is dependent on the goodwill and initiative of forest officials, who do recognise the need for a change in perspective, and on the ability of local people to trust these officials despite their different roles and power discrepancies. It is supposed to evolve through the existing management structure.

The review of joint forest management experiences outside protected areas pointed out some of the successes and deficiencies in this approach. It was effective where individual forest officials had the flexibility and initiative to chart a creative course, and problematic in organisational contexts where the philosophy of partnership was articulated but not really practised. The particular characteristics of protected areas in fact, make them less conducive to joint forest management approaches. There is a higher level of Forest Department authority, a more centralised management approach, and a higher level of mistrust between authorities and local people. There is less freedom for the creative and innovative forest official. The fact that the recent joint forest management guidelines do not apply to protected areas (SPWD, 1993) is evidence that Forest Departments do not view this approach as appropriate in these settings, and are unlikely to take the initiative required of them to change the guidelines and take action.

JPAM has appropriate goals which provide an alternative to the present model, and it deserves testing in an appropriate setting with innovative officials and low levels of conflict and mistrust. Yet we suspect that it would only speak of a new philosophy while leaving past management practices unchanged. We believe it is only the exceptional person, be it forest official or local inhabitant, who voluntarily gives up personal powers and authority while still having the mandated responsibility for the decisions based on those powers. We do not believe that Forest Departments would volunteer to give up their strong centralised authority in protected areas, but this is required in the JPAM partnership approach.

The Rajaji area again provides a case study illustrating the problems with a JPAM approach. It is impossible to visualise, after the past fifteen years of conflict and the enormous discrepancy in powers, forest officials and local people sitting around the table in equity and partnership under the structural framework of the present system. The Forest Department has an unlimited authority to determine the economic survival and place of residence of a Van Gujjar. The Van Gujjar is paying up to a third of the family income in fees and taxes, to officials in the same organisation with whom the Van Gujjar is to be an equal partner, and this figure can go up or down arbitrarily. These types of problems have been documented all over India in a wide range of protected areas (SPRIA & RLEK, 1993). The JPAM concept of an honest partnership under the existing system is not a viable option in many settings. There must be a structural change in powers and systems in order to facilitate an effective community management model.

Definition and Rationale for Community Forest Management in Protected Areas

We have utilised the wealth of available knowledge, and experience with alternatives for local involvement (see Part I), in order to define community forest management in protected areas as:

Community Forest Management in Protected Areas (CFM-PA) is a set of organisational structures and processes for defining and managing protected areas, in which the local peoples who have the traditional rights to inhabit and/or use the area become the leaders in managing the resource, while government departments and other stakeholders have a monitoring and support role. The overall goal is to protect and administer in a sustainable manner the ecosystem, its wildlife, and traditional rights and lifestyles of the local peoples integral to it.

This definition requires a clear change in the roles, structures and processes of protected area management, and defines local people with traditional lifestyles as integral to it. A wide range of questions must be answered in each protected area, in order to implement the model based on the characteristics of the ecosystem and the local people who have traditionally used it. A detailed research process must be undertaken with local people in each setting to develop specific proposals for implementation. Yet this initial research can be a temporary process, as the traditional knowledge of local people would be drawn on through their leadership roles during implementation, while scientific methods of evaluation would monitor the process and results.

We believe there are many multiple reasons for favouring the CFM-PA approach:

1. *Effectiveness of Forest Protection:* Experience and research in local involvement outside protected areas has shown that the community approach is most effective in instances where one has capable communities or communities in which the appropriate management capacities can be developed. We believe this can apply to a wide range of communities. Moreover, CFM-PA clearly identifies the structural reasons for the failures of existing policies and management systems, and proposes specific alternatives to overcome those.

2. *Cost:* Community Forest Management in Protected Areas would inevitably cost government less, as it is based on local communities taking over government roles. The relatively low cost avoids subservience to large international donors, who in turn have specific vested interests which may run counter to the interests of the ecosystem or local people.

3. *Respect for the Rights and Needs of Indigenous People:* Community management is inherently more respectful of the rights of indigenous people, providing mechanisms through which traditional knowledge is given credibility and validity in the wider society.

4. *Insulation from Vested Interests:* Placing power in the hands of local people and transparent decision-making

structures provides an obstacle against present vested government and political interests overwhelming the reform process.

5. *Community Development and Environmental Consciousness:* The community management process itself could develop the capacity of the community which builds cultural pride and identity that in turn could be transferred to other contexts. It could provide a practical example of a viable alternative lifestyle to the dominant urban consumer mentality that is sweeping the world, and is a major element of environmental destruction. Protective area management could become an example of the positive environmental practices of alternative cultures, rather than a means to distract attention from the broader environmental and social problems of mainstream culture.

Some may view this approach as idealistic, given the present Forest Department management system for parks and protected areas in India. We believe however, that it is *more* realistic in that it requires a restructuring and redistribution of management roles at the outset if it is to succeed. It does not make the assumption that an institution, which has operated for 125 years with strong vested interests in one mode of operation, would awaken and change itself. This restructuring process would be difficult but it is much more effective to do it in an open and direct fashion, rather than attempt to convince officials to give up powers which they exercise in their own vested interests. Local peoples cannot participate in the existing system as equal partners, when the officials sitting across the table from them have powers over all aspects of their livelihood and lifestyle; either the style in which they build their homes or the amount they pay in fees to secure basic needs. Powers must be openly defined at the outset so as to eliminate arbitrary authority.

Our experiences and existing research suggest that there are creative and innovative forest officials who would welcome this process. It would be essential that they accept the premise of the model in theory as well as in practice. The model would not be implemented at the same time in a wide range of contexts through massive funding from an international donor. It would not bring large scale funding to national coffers for a limited time, with the ultimate expectation that the government would continue that funding. Instead, it makes long term economic sense because it requires the government to do less and pay less than for the present system which, despite the cost, is failing to protect the environment and the local people who depend on it. We believe that with careful discussion and advocacy, the government will see the wisdom of adopting this approach in experimental settings, given the present deficiencies and costs of the existing system. Prime Minister Rao in fact, endorsed this type of approach in principle in 1992.

Principles of Community Forest Management in Protected Areas

The principles for CFM-PA defined below are based on research into the effectiveness of community management strategies that were reviewed previously (see Part 1). There are ten principles which define the structures and processes of the approach. We expect these principles to be appropriate in all settings, though their implementation would be tailored to local needs.

Ecosystem Protection

It is essential that CFM-PA is structured to protect the ecosystem, the wildlife, and the rights and needs of indigenous people who are integral to it. Protected areas have generally provided the legitimate function of safeguarding forest areas from development by urban and commercial interests, and must continue to do so. Indigenous people who had the forest for subsistence have a specific interest in forest protection. Sustainable use of the ecosystem has been their traditional right, and must be recognised through their lead management responsibilities. The CFM-PA approach must ensure, through its structures and provisions, that commercial and urban interests do not exploit protected areas under the guise of local control.

Community Responsibility

CFM-PA redefines the distribution of authority and decision-making powers, investing local people with the lead responsibility for managing the protected area, while providing effective mechanisms for monitoring and support from Government agencies and other stakeholders. Forest dwellers are the most knowledgeable managers of the forest, who have a specific investment in using and protecting it in a sustainable manner, if they are to be assured of long term tenure. Those who live closest to the resource must have the greatest responsibility in managing it. Government does have a legitimate and important monitoring and support role.

Participatory, Democratic Structures

The approach provides for open and democratic management structures, which ensure maximum participation from all members of the community. A major deficiency at present is that individual officials can make decisions, based on their own self-interest, behind closed doors which run counter to the interests of the ecosystem and the local people. Yet under the existing system no-one is in a position to challenge their absolute authority. Inevitably there will be individual community members with own interests which favour over-exploitation, but open and participatory structures would block their ability to take control as over-exploitation runs counter to community interests.

Forest protection and use is a shared responsibility of all community members, and, all therefore, must participate. Wherever possible, decisions should to be made in open forums rather than by representatives who can exert influence over time to block wider participation. Where representatives are required, they must be democratically elected with a mechanism to recall them and hold them immediately accountable for their actions. Wherever possible, decision-making responsibility should be delegated to committees made from all members of the community who are involved in using and protecting the resource. Experience in other jurisdictions shows that the processes must be independent of party politics.

Effective Conflict Resolution

An open and participatory management system would inevitably produce discussion, debate, and conflict even within relatively homogeneous communities. There would be problems, and therefore, it is essential to build conflict resolution mechanisms into the framework. It is preferable to utilise traditional structures, if they have a history of effective use in the local community, or to develop new structures based on traditional approaches and practices. In this way local people would have a solid understanding and sense of ownership of the processes at the outset, so that they would be less vulnerable to manipulation by exploitive individuals either within or outside the area. The existing justice system is seen as corrupt and inaccessible by many local people. It does not resolve conflict effectively at community level. Some critics predict conflict under CFM-PA but open conflict with effective conflict resolution mechanisms is far more preferable to back room decisions involving a few powerful personages.

Open Communication

It is essential that there be many open communication channels for information, so that management issues and problems can be rapidly identified and resolved. Openness breeds accountability from all involved, as that exploitive activities and decisions would be known by all community members immediately.

Discrete Jurisdictions and Explicit Agreements

Individual communities or groups should be responsible for discrete areas with clear boundaries. It is preferable to create smaller management groups for smaller areas rather than create large areas managed by heterogeneous groups. There must be explicit agreements between all participants.on responsibilities, usage, benefit-sharing, and protection policies at the outset. Mechanisms for appeal and conflict resolution must be described in these agreements.

Traditional Rights and Use

Protected areas are designed to protect the environment and prevent exploitation. Current use must be in keeping with the traditional lifestyle through which indigenous people have been an integral part of the ecosystem. Commercial interests cannot claim rights to use protected areas under community management, nor can traditional communities change the nature of their use to satisfy their own commercial desires. This principle is in keeping with the Forest Policy of 1988, which emphasises the importance of forest resources for environmental protection, and meeting the subsistence needs of local communities, rather than satisfying commercial interests.

Responsibility and Benefit Sharing in Relation to Traditional Usage

Once an area is defined, all groups who use the area must take responsibility for it in relation to their type and level of traditional use. Groups who have a greater traditional usage and presence would have more responsibility than those who have less usage. Yet there must also be safeguards in structures to insure that the needs of all groups are protected, so that one cannot exclude the other. This same principle applies to power within groups, so that all individuals traditionally using an area must have responsibility for it such, that individuals cannot be excluded. In instances where many groups are traditionally using the same area for different purposes, there must be structures and incentives to allow for open communication and co-operation between them.

Gender Equity

Gender equity is an important principle closely related to that of responsibility and benefit-sharing in relation to usage. There is the need to ensure that women who have traditionally been important users of the forest are assured important roles in decision-making with regard to community forest management. Some community management experiments have been compromised, in that women as primary users were not involved in primary decision-making, so, that their needs were not met and they refused to support the system (Sarin, 1995).

Effective Monitoring and Advocacy

There must be effective and open mechanisms for monitoring and evaluation built into all aspects of the structure and process. Sustainable development has been defined as a set of structures that "learns fast from its mistakes in the use of natural resources and rapidly rectifies its human--nature relationships in accordance with the knowledge it has gained (Agarwal & Narain, 1992, p. 160). All communities will make mistakes, and the key is to learn from them and adjust. Effective evaluation as an on going process is essential. It is even more important in experimental programs so that maximum learning can result which can be applied to new contexts. Monitoring must come through a multi-layered sectoral committee that represents all those with legal and traditional responsibilities for the area under community management.

The regional committee also provides an organisational structure which can mobilise support from many sectors to advocate for effective policy implementation in line with community management priorities. One of the difficulties with the present top-down bureaucratic system is that well-intentioned officials cannot affect policy above their levels and hence are powerless in their roles. On the other hand, exploitation interests outside the system can gain access at many levels. It is essential to utilise well-motivated and influential individuals and organisations to advocate on behalf of community management priorities. The structures for monitoring and advocacy must be transparent, open, and accountable to the community and their representative organisations for their positions and work.

These principles are interconnected and complementary such that the failure to incorporate any one of them into specific community management proposals would threaten the integrity of the whole process. Thus participatory decision-making structures must go hand in hand with ecosystem protection to assure that the needs of the ecosystem and the people are both met. Community responsibility could produce incurable conflict if there are not discrete jurisdictions, explicit agreements, and responsibilities and benefit sharing in relation to usage. The combination of principles into an overall approach would result in maximum incentives for forest protection.

It is noteworthy that the emphasis in CFM-PA, as it applies through all settings, is on the processes and structures to be utilised, rather than on the goals, objectives, and strategies through which to protect the ecosystem and the rights of indigenous people. In fact, the objectives and strategies would vary greatly depending on the local context. Strategies which would work well in one locale might be unreasonable in another setting. This emphasis on process and structure over content is in contrast to the present system which emphasises consistent objectives and strategies across all protected areas. Common strategies are implemented by a standardised system that leaves no opportunity for diversity, flexibility, experimentation. It is not coincidental that the present approach is hostile to the definition of sustainable development as the ability of social structures to learn from their mistakes. The present management system has not learned from its mistakes, and has not achieved the sustainable development of protected areas, even in the traditional environmental definition of the term. Approval of an experimental context in which to test CFM-PA would be a bold step to open up discussion and create both structural and environmental sustainable development opportunities for protected areas.

Community Characteristics for Experimentation with CFM-PA

Experimentation with CFM-PA should begin in settings which seem conducive to success, based on research and experience in others outside protected areas. Lessons from these locales could then be applied in a broader range of settings. The key community characteristics identified from the research on community forest management in chapter 2 are:

- a culturally homogeneous make-up and set of values
- high percentage of tribal people
- lower levels of economic disparity

- high levels of dependence on the forest for survival
- clearly defined forest use areas with minimum overlap with other groups and interests
- greater gender equity and involvement of women
- land with capacity for regeneration rather than fully degraded land where success would be slow and small

The Rajaji Area as a Test Case for CFM-PA

In the Rajaji context the Van Gujjars, who are the primary users of the core area, particularly exhibit the above community characteristics. The Van Gujjars:

- are a culturally homogeneous tribal group with a clear and strong set of traditional mores that value the forest, wildlife, and ecological protection.
- are entirely dependent on the forest for their survival and prosperity, so they have an inherent need to protect it.
- have clearly defined areas of forest use and zones of control, both as a community relative to other communities, and as individual families within the community.
- respect the traditional rights of other local people to use forest resources and are willing to co-operate with them.
- have the extensive indigenous ecosystem knowledge and expertise, accumulated over generations as forest dwellers, which is required for successful management.
- have traditional and effective community decision-making structures, which operate by consensus and have not been corrupted by political parties.
- occupy lands with the capacity for regeneration, rather than areas where such efforts can bring little or slow returns.
- live in close proximity to each other, are fully aware of each others' activities, and are highly motivated to gain respect from peers, thus peer pressure is a powerful means of social regulation.
- have relatively low levels of economic disparity within the community. Economic differences are less prominent than

the shared cultural values which bind Van Gujjars to each other.

- have greater gender equality than in many rural cultures.
- are highly motivated, and requesting that they be given the opportunity to manage the forests.

According to Talib Pradan, Van Gujjar of Kaluwala Khol, "We get married in the jungle, our children are born in the jungle and we live our whole life in the jungle. People in the towns, in the cities think about what their things cost, but we are more concerned if anybody should cut our trees. If somebody from outside comes and cuts our trees we will not be able to say something directly to them, but inwardly our hearts will bleed. The Van Gujjars in this position are at a disadvantage, because they do not have the power to say anything to the person who comes from outside. If the people from the organisation and the government co-operate with us, then we can save the forest. In this arrangement the forest will keep growing. People say that there is value worth crores in the national park, and so much money has already been spent on it but the results have not been seen. But if the money is spent on a marginalised community the results would have been very visible, and they would have prospered."

For these reasons, we believe that the Van Gujjars are the ideal community to take lead management of the core area as a test for CFM-PA within the Rajaji context. CFM-PA also means giving lead responsibilities to local villagers in relation to the resources and areas where they are the primary users. They also exhibit many of these characteristics although less clear cut in relation to those who use the proposed park area and resources.

- They are generally subsistence people with a high level of dependence on specific forest resources. The most important user group, the baan workers, have nearly complete economic dependence on the bhabbar grass resource.
- There is relatively less economic disparity as most villages on the periphery of the proposed park are poor.

- In most areas, particularly in the proposed pilot zones, villages have separate areas of usage with boundaries based on tradition and proximity. There is little overlap with the Van Gujjars, and there is clarity in traditional uses with respect for each other's rights where there is overlap.
- They use lands with the capacity for regeneration, rather than areas where regenerative efforts can bring little or slow returns.
- They are motivated to take responsibility for management, and are explicitly requesting involvement, particularly in the initial pilot zones. The Taungya villagers, in particular, have a high level of indigenous knowledge, and desire to take leadership.

There are more difficulties in implementing CFM-PA with local villagers than the Van Gujjars, yet overall the context provides an excellent test case, particularly in the initial pilot zones where villagers have a particularly high dependence on bhabbar grass, low economic disparity, fewer cultural differences, and a high level of motivation. The villagers also have a smaller role in using the proposed park than the Van Gujjars. Their primarily used areas are on the proposed park borders. There are large numbers of villages with more complexity and overlap in the Gohri range, which is why this area does not come under CFM-PA until the final phase in the Rajaji area. There are also elements in some villages, responsible for illegal poaching and felling in the proposed park area, which may cause conflict for Van Gujjars and villagers. Effective community enforcement structures and powers would be essential to address this issue.

Overall, community forest management in the Rajaji area represents an excellent test case for CFM-PA. It is the best chance of protection of a threatened ecosystem, and the support of the rights and lifestyles of marginalized local people. The next section describes detailed objectives, strategies, structures, and processes for the implementation of this approach in the Rajaji area.

A Brief Look at
Community Forest Management in Protected Areas
(CFM-PA)

The Approach...

CFM-PA is a set of organisational structures and processes for defining and managing protected areas, in which the local people who have traditional rights to inhabit and/or use the area become the leaders in managing the resource, while government departments and other stakeholders have a monitoring and support role. The overall goal is to protect and administer in a sustainable manner the ecosystem, its wildlife, and the traditional rights and lifestyles of the local peoples integral to it. CFM-PA requires a clear change in the roles, structures and processes of protected area management and defines local peoples with traditional lifestyles as integral to it.

The Rationale...

The existing protected area management system has failed to protect the ecosystem and the indigenous peoples who depend on it. The involvement of local people is essential. There are many reasons in favouring for the CFM-PA approach:

- Effective Forest Protection
- Low Cost

- Respect for Rights & Needs of Indigenous People
- Protection from Vested Interests

- Community Development and Environmental Consciousness

The Principles...

Ten principles are to be implemented in relation to the characteristics and needs of the local communities and ecosystems...

- Ecosystem Protection
- Community Responsibility

- Participatory, Democratic Structures
- Effective Conflict Resolution

- Open Communication
- Traditional Rights and Use

- Management Responsibility Benefit Sharing in Relation to Traditional Usage
- Discrete and Jurisdictions and Explicit Agreements

- Gender Equity
- Effective Monitoring and Advocacy

Chapter 10
Detailed CFM-PA Proposals for the Rajaji Area

The Rajaji Context for Community Forest Management

Before proceeding with specific proposals, it is important to consider two elements that are beyond the scope of this plan, yet are important to the long-term success of community forest management in the Rajaji context.

Firstly, there is an urgent need to reform the Wildlife Act in its provisions on the rights of local people (Ramesh, 1996). This would include revisions to recognise and protect the traditional rights of local people, to create more flexible categories of protected areas, and to grant management authority to local bodies (Law Faculty of Delhi, 1995; Ramesh, 1996). Positive proposals for law reform have already been initiated in several forums (Kothari, 1995b). The Government of India has recognised the need for reforming the Wildlife Act, and has constituted an advisory committee to consult the public and recommend new approaches. Law reform however, is always a slow process and the present efforts face the additional obstacle of having to start anew with the Centre Government which comes to power after elections in 1996.

Law reform is important in the long term but it is beyond the immediate scope of this plan. In the interim, the plan can be implemented through administrative discretion, in that the Rajaji area has not been proclaimed a national park because of the outstanding rights. As a result it must be managed under Wildlife Act provisions relating to sanctuaries, which leave room for administrative discretion to continue traditional rights, and delegate management responsibilities through the park warden. The state and central governments should see the wisdom of this approach, considering the failure of the present system, the constitutional support for this approach, the particular qualities of these peoples for community management, and the need for experimentation while law reform proceeds. Law reform is important however, in the long term, as there is no guarantee that a government which uses its discretion to sanction a policy will not also withdraw that approval at any time.

A second long-term issue is the need for Van Gujjars, with the help of NGOs and government, to develop alternative economic options for those who choose to leave the forest. The increasing Van Gujjar population cannot wholly be supported in the long term in the core area forest, while some Van Gujjars may want to make alternative choices. Better educational opportunities, which are built into this plan, are one means to this end. At present, the only alternative economic option that has been available is the notion of moving to some agricultural land in general and rehabilitation to Pathri in particular. We believe this is one long term option that can be offered to Van Gujjars by the government on a truly voluntary basis. This option however, given the present mistrust and bad faith is not voluntary, but based more on a wide range of threats and pressures. As the recent events in Pathri have demonstrated, the place has been utilised only by a few Van Gujjars with nothing to lose. Although some Van Gujjars are requesting agricultural land, they reject Pathri under present conditions and based on its previous record do not believe that the government will negotiate with them in good faith. Moreover, the pressure on the ecosystem is not reduced by Van Gujjars moving to agricultural land under the present management system, as there is always the option to pay fees to maintain one's buffaloes in the forest. We suggest two steps to be taken prior to further consideration of the land option:

1. The government should follow through on unfulfilled promises to those who have already occupied land, including people who moved out of the proposed park area in the 1970s, and those who moved to Pathri. These promises include provision of official title to the land, electricity, irrigation, and school facilities.

2. Implementation of the community forest management plan should begin so that Van Gujjars have an alternative to the existing system. They must also realise that the land choice

would no longer include permission to leave buffaloes in the forest.

Even Van Gujjars from Ranipur and Motichur who live in the poorest forest, and have more frequently voiced the desire for agricultural land, despite their recognition of the difficulties, have been supportive of community forest management and eager to participate. Only once the new system is in place would the true nature of their choices be evident. In the meantime, the government could show good faith with respect to past land offers, which would in turn lay a framework for more successful and honest offers to Van Gujjars who want to leave the forest in the future.

Both legal reform and viable economic alternatives for Van Gujjars in the core area are beyond the scope of this plan, and must be achieved by a range of constituencies on a long term basis.

Objectives of Community Forest Management

1. To protect the ecosystem of the Shiwaliks, conserve biodiversity, and protect and support endangered and threatened species.

2. To protect and support the traditional rights of the Van Gujjars, so that they can live permanently in the protected area, in an environmentally and economically sustainable manner.

3. To protect and support the traditional rights of villagers living in border areas to use minor forest resources within the protected area in an environmentally sustainable manner.

4. To increase documented knowledge about the Shiwalik ecosystem, through a partnership between local people and external scientists and researchers.

5. To ensure and promote ecologically and culturally responsible tourism and education, under the supervision of the community forest management structure.

Broad Strategies for Achieving Objectives

The plan objectives are to be addressed through the following broad strategies, which are subsequently presented in detail :

1. *Core Area Community Forest Management Structure:* Development of a community forest management structure for the core area, in which the Van Gujjars have lead management responsibility in co-operation with local villagers, with monitoring and support from existing agencies such as the Forest Department, RLEK, and police officials. A formal management agreement and the Regional Committee would establish the policy framework.

2. *Villager Minor Forest Produce Committee Structure:* Development of a Minor Forest Produce Committee (MFPC) system for local villagers to manage the forest resources and border areas which they have traditionally utilised, in co-operation with Van Gujjars, and with monitoring and support from existing agencies such as the Forest Department, RLEK, and police officials. A formal management agreement and the Regional Committee would establish the policy framework.

3. *Community Forest Protection Structure:* Development of a community forest protection system for the area, in which Van Gujjars are trained to take on front-line protection duties with back-up support from the police and forest officials.

4. *Support for the Van Gujjar Nomadic Lifestyle:* Advocacy and support for the nomadic movement of Van Gujjars, so as to maintain the viability of this lifestyle for those who choose it, and reducing the impact of Van Gujjars on environmental resources in the summer, to allow the Rajaji forests to regenerate.

5. *Van Gujjar Development Priorities:* Assisted Support for Van Gujjar requests for development priorities, providing programs in which they take responsibility for policy direction and implementation, particularly in the areas of education, health, and milk marketing.

Designation of Areas

Map 7 (see Chapter 5) divides the proposed park area in relation to the uses of local people based on extensive field interviews and discussion with Van Gujjars and local villagers. There are two categories of land use by local people.

1. *Core Areas:* Areas where Van Gujjars are residents and the primary users, with the exception of villager extraction of bhabbar grass and occasional grazing and fuelwood collection. This includes most of the park area, particularly the inner areas.

2. *Border Areas:* Areas where villagers collect minor forest produce and are the primary users of the forest, although there may be some lopping done by Van Gujjars. Van Gujjars do not reside in these areas which are on the periphery of the proposed park.

Co-operation between local peoples is required to manage both areas, but differences in structure and leadership are proposed for the two types of areas, based on the differences in traditional usage, benefits, and responsibilities.

Proposals for Core Area Community Forest Management Structure

The Van Gujjars are proposed for the lead managers of these core areas, working in co-operation with the villagers who would take responsibility for the bhabbar resource they utilise in the area. Administrative and management responsibilities would be delegated through the policy and legal framework of the community management agreement and the Regional Committee. The Van Gujjars have lead management responsibilities in these areas because they are the only residents and forest dwellers, and have a high level of dependence on these forests for all aspects of their survival. They do not use the bhabbar grass, but respect the traditional rights of villagers to extract this and to take responsibility for management of this resource. In some khols there is no villager use, while in others there is significant extraction of bhabbar grass, some grazing, and fuelwood collection. The core area management structure provides for villager roles in relation to their level of use. The pradhan of the relevant village panchayat is a member of the decision-making khol committee.

Figure 1

Functions for the proposed core area community management structures.

Primary Functions & Level	Level Community Management Structures	
Decision-Making/ Khol Level	Khol Committee	Minor Forest Produce Committee
Mediation & Conflict Resolution/ Range Level	Range Panchayat	
Co-ordination Sanctuary Level	Sanctuary Committee	
Policy Framework Monitoring and Advocacy/ Regional Level	Regional Committee	

The management agreement would restrict use of the forest resources to minor forest products traditionally used by Van Gujjars or villagers. It would require that the on-going use of these products must be in keeping with the traditional lifestyle of Van Gujjars and local villagers. Commercial interests cannot claim rights to use the protected area under community management, nor can Van Gujjars or villagers change the nature of their use for new commercial purposes.

Figure 1 provides an outline of the proposed community management structure for the core area, indicating the decentralised nature of the administrative structure. The implementation work is at the family and dera level, decision-making within the overall policy and legal framework rests at the khol committee/MFPC level, conflict resolution for internal community matters is dealt with at the range level, overall co-ordination is at the sanctuary level, while the policy framework is set and monitored at the regional level.

Management Agreements

Management areas would generally be defined in relation to the geographic boundaries of khols assuming approximately twenty Van Gujjar families per area. There would be a detailed written agreement signed between the Forest Department, all members of the khol committee, and the NGO (RLEK). This agreement would establish the overall legal and policy framework and committees would function within this delegated mandate. The framework would be set according to this plan and through the Regional Committee.

There could be two khol committees in very large khols or one committee for two khols in areas with few Van Gujjars. The exact breakdown of the area of each committee is defined for the proposed pilot zone in Chapter 11. In the event that a Van Gujjar family does not want to participate in community management, and we have not met one to date, that dera would have to shift to another area not under community management with an alternative family moving into the area. All families in an area must adopt community management to make it effective. The agreement would cover a five year period and lay out the rights and responsibilities of all parties in accordance with the outline defined below.

Where applicable, a second detailed written agreement would be signed by members of the villager Minor Forest Produce Committee (MFPC) managing the bhabbar resource in the area, the khol committee responsible for working with them, the Forest Department, and the NGO. This agreement would cover the same five year period and lay out the policy mandate, and the rights and responsibilities of all parties as described under the village Minor Forest Produce Committee structure later in this chapter. The MFPCs are defined for the pilot zone in chapter 11.

Dera Proposals

Each dera would be responsible for developing a specific proposal for the next five years for the number of buffalo to be supported within their forest area, the actions they would take to improve the forest, and the assistance they would like in forest rejuvenation. The proposal would initially cover the area of their existing permit under the present forest management system. If deras are now jointly sharing a permit, they would have the option to sub-divide their area by dera for the purposes of community management. The dera proposal would include topics such as:

- Description of the forest and estimation of the buffalo capacity.
- Plans for tree planting including the number, species, timing, assistance required etc.
- Lopping frequency required to for the number of buffalo given the forest quality and quantity
- Plans for the management and reduction of weeds
- Plans for the summer season, number of people and animals remaining in the khol
- Specific proposals and assistance required for supporting wildlife, providing suitable water, maintaining firelines, etc.

Deras would require assistance in developing these proposals, particularly with regard to literacy issues. This would be provided by the training facilitator and/or the DFO forester for the sanctuary. Proposals would be presented to the full khol committee by each dera. Once the khol committee approves the proposal, possibly with revisions, the dera would implement it and report on progress to the khol committee every six months. The dera could propose revisions to the plan for approval by the khol committee.

A dera could not increase the size of its herd beyond the maximum size approved by the khol committee in the dera's proposal. The khol committee would enforce these limits through its peer oriented discipline structure.

Khol Committee

(a) Membership and Direction: Each dera must have one member on the Khol Committee, but however, many adults that choose from the dera may participate. One-third of the members must be women. The Committee would decide, in an open meeting by consensus, with a person to facilitate committee meetings, and a person to record decisions. The training facilitator and the DFO forester for the range would be participant observers in the Khol Committee, but would not have a vote in determining decisions made by the Committee. If there are villagers extracting bhabbar grass in the Khol, two representatives of the villager MFPC responsible for managing this resource would participate in all Khol Committee meetings. One village representative would be the pradhan of the relevant village panchayat. The villager roles would be related to issues which affect the minor forest produce that they utilise in the area, principally bhabbar grass. The Khol Committee would meet as frequently as it saw fit, with but a minimum of four times per year.

(b) Decision-Making: Decisions are to be made by consensus of all members present and meetings must be rescheduled if a key decision is to be taken of concern regarding an absent member. If the committee cannot come to agree with respect to a decision, it would be referred to the Range Panchayat for arbitration and a decision. All deras would, as part of the initial CFM-PA agreement, abide by the decisions of the Range Panchayat, in the event of their involvement in a khol dispute which is referred to the panchayat.

(c) Khol Carrying Capacity: The Khol Committee would accept the recommendation of the Range Panchayat (see below) on the carrying capacity of the khol, and the number of buffalo permitted in the khol in the winter and summer seasons. The carrying capacity must be equal to the number of buffalo permitted in the Khol in each season, unless the Khol Committee can show the panchayat that they would undertake forest improvements to compensate for any discrepancy (tree planting, purchase of fodder, etc.). If there are too many buffalo in the khol, the Range Panchayat would

work with the Khol Committee to shift them to another area. If the Khol carrying capacity is higher than the number of buffalo in the Khol, additional buffalo may be shifted from a Khol which has over its carrying capacity.

(d) Appeal against Decisions: The Khol Committee could appeal to the Sanctuary Committee against the decision of the Range Panchayat on the carrying capacity if they find it unacceptable. In such as instance, the Sanctuary Committee would review the decision of the Range Panchayat. If the Sanctuary Committee agrees that the appeal is legitimate, it would request the panchayat from another range to do a second assessment and make a final decision.

(e) Khol Committee Responsibilities: The Khol Cmmittee would be the key administrator of activities in the Khol, and would function within the overall policy framework and legal structure, mandated under community management, and monitored by the Regional Committee. The Khol Committee approves individual dera plans, and would have the ability to re-apportion forest within the khol in order to match the number of buffalo with the forest as necessary. The Khol Committee would work by consensus, which means that no decisions would be made unless all members agree. The re-apportioning of forest could occur only once at the start of the agreement, and would be permanent thereafter. The provision for re-apportioning the forest at this time is based on the fact that traditional permit areas have been shifted by forest officials during recent years in a manner not necessarily in the interests of forest protection or equity to Van Gujjars. This represents opportunity to adjust forests where necessary, although it is expected that the apportion would follow existing patterns. The approved dera plans would replace the existing forest permit system, as the means of regulating forest use.

Disputes in the Khol Committee would be judged by the Range Panchayat, but there would inevitably be pressure to solve matters internally without calling in the panchayat. The essence of community management is to place day to day responsibility at grassroots level, while still maintaining dispute mechanisms and means to insure that local self-interest does not outweigh other concerns. The Khol Committee would also provide a forum for planning and supporting each dera's efforts to improve the forest. Development or conservation projects could not flourish in the khol without the support and approval of the Khol Committee.

If bhabbar grass is extracted from the Khol, the Khol Committee would be responsible for reviewing and endorsing the management plan proposed by the Village Minor Forest Produce committee responsible. If the two Committees could not agree on the plan, the dispute would be referred to the Range Panchayat for resolution. Khol Committees would recognise the traditional grazing and fuelwood collection rights of villagers in the khol.

A final major Khol Committee responsibility would be the definition of plans to insure an effective community presence in the areas in the summer months when Van Gujjars move to the hills. Community forest guards (see below) would remain in the area during the summer months, as well as the other people who traditionally have remained behind to protect deras and watch over the forest. This is not a departure from recent years, when some Van Gujjars have remained behind to fulfil these functions in an informal way. Fifteen or twenty Van Gujjars remaining behind to watch over a khol would represent an effective and realistic presence in a khol. Buffalo would move out of the area in line with past practices and carrying capacities. Table 12 gives a complete list of khol committee responsibilities.

Table 12:
Responsibilities of the Khol Committee

The committee is responsible for ...

Forest Improvement
- To make decisions by consensus on the individual forest management proposals from each dera. The committee could re-apportion the forest within its jurisdiction in line with the needs of the individual deras and the number of

buffalo they own. The goal would be to assure that the forest for each family is sufficient for the number of buffalo.

- To propose and carry out projects for forest improvement, wildlife protection, soil erosion, or water management projects in the khol.
- To work co-operatively with the village MFPC by selecting two Van Gujjars to attend MFPC meetings regularly.
- To review and endorse the village MFPC management plan for bhabbar grass for the khol in a joint meeting with the MFPC. Disagreements over the plan would be referred to the Range Panchayat.

Forest Protection
- To advise and assist deras in making plans to improve the quality and carrying capacity of their forest, making sure that the number of buffalo held by a family would be in accordance with the permitted number approved in the dera plan by the khol committee.
- To supervise the work of the Van Gujjar community forest guards in the khol, and ensure that there is an adequate presence of Van Gujjars in the khol for supervision purposes in the summer.
- To develop a plan to fight fires in the khol, or to assist other khols' request for help.
- To report illegal activities to the appropriate authorities, and to report to the Regional Committee if the authorities fail to follow-up satisfactorily on the reports.
- To determine disciplinary action against people in the khol who break khol committee policies and practices.

Accountability and Reporting
- To refer disputes within the committee to arbitration by the Range Panchayat. If it were determined that buffalo must leave the khol due to a low carrying capacity, the khol committee would make the decision by consensus as to which buffalo must move elsewhere.
- To function within the community management agreement and the policy framework established by the Regional Committee.

- To work with the DFO forester to provide a forest inventory, at the start of the agreement, and another at its conclusion (see section 10.5.7).
- To report regularly to the Range Panchayat and Sanctuary Committee on the problems and the accomplishments under the community management structure, seeking their assistance where appropriate.

Range Panchayat

The Range Panchayat would be the essential co-ordinating body for resolving disputes in the community management structure. It is based on the traditional system of respect for the decisions of a council of elders. It is essential to have honest, wise and respected people on the panchayat, and both Van Gujjars and villagers could accurately identify these people, assuming an open, consensus building appointment process. The panchayat would also play a crucial role in determining khol carrying capacity.

(a) Appointment and Tenure: A Range Panchayat would consist of a minimum of five Van Gujjars including one from each khol, and one representative from each MFPC in the range. For the Van Gujjars, each khol would nominate a member from its own khol, and the list of nominees would then be reviewed by every khol committee. Final decisions would be made at a range meeting so that every committee agrees to all members of the panchayat. For the village members, each MFPC would nominate a member, and the list of nominees would then be reviewed by every MFPC using the range. The executive of every MFPC would then have to agree to the final villager list for the panchayat. The panchayat would serve for three years, when a similar nomination process would begin. The panchayat could be reviewed at any time prior to that, if a majority of the khol committees and MFPCs request the Regional Committee for such a review.

Table 13
Responsibilities of the Range Panchayat

The Range Panchayat would be responsible... for :-

Khol Committee and MFPC Relationships

- To assess the carrying capacity for each khol in the range, and approve khol proposals for the total number of buffalo allowed in the khol in each season.
- To arbitrate and resolve disputes within khol committees and MFPCs when requested to do so.
- To act as a co-ordinating body within the range to shift buffalo across khols if required.
- To assist and support khol and MFPC committees in their work.
- To supervise the administration of funds for khol improvement projects.

Sanctuary and Regional Committee Relationships

- To make recommendations on the status of closed khols in the range to the Sanctuary Committee.
- To provide recommendations on overall forest regulations to the Sanctuary Committee after consultation with the khol committees.
- To work with the Sanctuary Committee in its efforts to co-ordinate activities across ranges.
- To function within the community management agreement and the policy framework established by the Regional Committee.
- To regularly report to the Sanctuary Committee and Regional Committee on the problems and accomplishments under the community management structure, seeking their assistance where appropriate.

(b). Responsibilities and Composition for Decision-Making: The specific composition of the panchayat for bearing a dispute would depend on the nature of the issue. If the issue involves core area management, and does not deal with villager concerns regarding minor forest produce, only the Van Gujjar members would mediate. If the issue involves only a border

area, where there is no Van Gujjar presence or concern, then the villager members would mediate. If the interests of both communities are involved, all members of the panchayat would sit together. The Van Gujjar members would determine the khol carrying capacity for the Van Gujjars. Table 13 lists the responsibilities of the Range Panchayat.

Sanctuary Committee

A committee would be established to co-ordinate activities between ranges for each of the three sanctuaries (Rajaji, Motichur, and Chilla). The Sanctuary Committee would include two Van Gujjar and two village MFPC representatives from each range within the given sanctuary. These representatives would be chosen by the consensus of the range panchayats within the sanctuary, and follow the same three year terms as those of the panchayats. Additionally, one range training facilitator and the DFO forester from the sanctuary would be appointed by the Regional Committee.

The Sanctuary Committee would be responsible to ensure policy co-ordination across ranges, according to the policy framework established by the community management agreement and the Regional Committee. It would be responsible for co-ordinating and shifting buffalo where ranges are exceeding their allotted numbers. They would also be responsible for supervising a tree nursery in the sanctuary to grow seedlings for replanting in the khols.

Table 14
Responsibilities of the Sanctuary Committee

The Sanctuary Committee would be responsible for:-

Co-ordination
- To co-ordinate communication and implementation of policy throughout the Sanctuary.
- To arbitrate, and resolve disputes, within Range Panchayats, when requested to do so.

- To consider appeals of khol carrying capacity when the khol committee objects to the Range Panchayat decision, recommending a second khol assessment if appropriate.
- To act as a co-ordinating body for shifting buffalo across ranges if required.
- To assist and support khol committees, MFPCs, and Range Panchayats in their work.
- To define forest regulations (e.g., lopping practices) within the policy framework of the Regional Committee, which would be implemented throughout the sanctuary after consultation with the Range Panchayats.
- To review and approve proposals from khol committees for forest improvement projects.
- To review and approve forest inventory procedures to be conducted under the DFO forester.
- To supervise the tree nursery.

Regional Committee Relationships

- To make decisions on the current status of closed khols in the sanctuary. If they are opened, only buffalo who are in excess of the carrying capacity in other khols or ranges may be shifted there.
- To report on expenditure for forest improvement projects of khol committees.
- To review and approve projects to be undertaken by external researchers in the sanctuary.
- To review and approve all eco-tourism and education proposals to be carried out in the sanctuary.
- To function within the community management agreement, and the policy framework established by the Regional Committee.
- To report regularly to the Regional Committee on the problems and accomplishments under the community management structure, seeking assistance and mediation where appropriate.

Regional Committee

The Regional Committee would be responsible for supervising the overall implementation of the community management structure and the policy framework. It would assure that there is open communication and healthy problem solving between various groups. It would not be responsible for day to day decisions, but for insuring that this proceeds in a healthy manner, and that it adheres to the policy framework as laid down in the management agreements. In addition, the committee would be required to adopt an advocacy role vis a vis the government, supporting the overall community management structure. The Regional Committee would receive a forest improvement budget from the Forest Department for forest projects proposed by the khol committees (e.g., soil erosion barriers, fencing, etc.). The Regional Committee would receive a training budget from the NGO (RLEK) for the initial training component to implement the community management structure. This would include funds to hire an individual responsible for evaluation and documentation of the community management process, who would report directly to the Regional Committee. Table 15 provides a detailed list of responsibilities.

(a)   Composition, Appointment and Tenure of Regional Committee:
The committee should be appointed by the Chief Conservator of Forests for Uttar Pradesh and should include:

- Three Van Gujjars, one from each sanctuary
- Three representatives from villager minor forest produce committees, one from each sanctuary
- The Director of each sanctuary
- The Conservator for the Shiwalik Forest Division
- The Chairman of RLEK
- Representative from the Wildlife Institute of India
- District Police Superintendent
- Three eminent people, including one Advocate without no vested interests in the Rajaji area. One each from Haridwar, Dehra Dun and Saharanpur Districts.

Recommendations for the Van Gujjar representatives would be provided to the Chief Conservator based on the consensus of open sanctuary meetings of Van Gujjars. The village representatives would be recommended by a meeting of the executives of village minor forest produce committees in the sanctuary. There would temporarily be more than one

Van Gujjar and villager representative from a Sanctuary in the early phases when only one or two sanctuaries are under community management.

The remaining Regional Committee representatives would be recommended by the consensus of the Sanctuary Director, Shiwalik Conservator, and the Chairman of RLEK. ·Appointment to this committee would be for a two-year period, it would be reviewed by each group, recommending either re-appointment of present incumbents or suggesting new appointments. Van Gujjar representatives could be reviewed at any time if requested by the Range Panchayats.

Table 15
Responsibilities of the Regional Committee

Policy Framework Establishment, Monitoring and Reporting

- To establish overall policy and the framework for administering it, in accordance with the community management agreement.
- To insure that all elements of the community management structure are performing within existing legislation and the policy framework.
- To monitor the progress of the community management structure by receiving reports from the Range Panchayat, Sanctuary Committees, the training facilitators, and the DFOs. These reports would include full financial statements from the Sanctuary Committees for the collection and expenditure of funds under Sanctuary Committee supervision.
- To supervise the expenditure on forest improvement and training component funds.
- To make a full public report to the Chief Conservator of Forests once a year on the progress and results of the project.
- To monitor the process and outcome of the full forest inventory, supervised by the DFO forester, working with the khol committees in the first and final year of the agreement.

- To determine whether there is a need for interim forest assessments, and recommend them in a particular area if appropriate.

Advocacy

- To follow-up and take action on complaints from the khol committees, MFPCs or the Range Panchayats concerning difficulties with law enforcement authorities, i.e., inadequate backing support to the khols, failure of authorities to proceed with legitimate legal action in the courts, etc.
- To advocate to the government, on behalf of the community management structure for the implementation of broader policies compatible with effective community forest management.

Review and Arbitration

- To provide support and arbitrate in difficulties within the Sanctuary Committees.
- To provide consultation on, and approval of all proposals from external people for research in the area.
- To provide consultation on, and approval of all eco-tourism and education proposals for activities in the area.
- To review and approve the procedures for the forest inventory, to be conducted under the auspices of the DFO.

Personnel Supervision

- To inform the Sanctuary Director, Shiwalik Conservator, and RLEK Chairman on the selection of staff whom these supervisors are proposing to employ under the community management structure within their organisations. Objections by the committee to individual appointments would be heeded by the supervisors and alternative individuals assigned.
- To review complaints about the performance of people working under the community management structure, where complaints have already been addressed directly to the supervisor concerned, and there has been no satisfaction, (i.e., khol committee complaints would first go to Range Panchayats, forest official complaints to the

Director or Conservator, training component staff to the RLEK Chairman, etc.).

- To select and supervise a member of staff with responsibility for evaluation and documentation of the community management process.

DFO Forester and Forest Inventory

One forest official at the DFO level in each sanctuary would be responsible for providing technical support and assistance to the khol committees and Range Panchayats during the course of the agreement. This person would function as a resource person/observer and would not have enforcement powers. The forester would be a permanent member of the Sanctuary Committee. He or she would be responsible for the leadership of full forest assessments in each khol, at the start of the plan and after five years, with smaller interim assessments if necessary. The assessments would be carried out through teams of Van Gujjars, using standard census practices. Additional technical assistance from forest department staff would be utilised if there were tasks for which Van Gujjars could not be taught the required skills within a reasonable period. The DFO would submit the procedures to be used in conducting the inventory for approval to the Regional Committee and Sanctuary Committee before starting the inventory. The DFO would be responsible for training the Van Gujjars to carry out the inventory as required. The results of the survey are to be reported in terms that could be understood by all participants in the community management structure. The Forester is responsible for regular reports to the Regional Committee on the progress of his or her work.

(a) Appointment: This person would be appointed and supervised by the Sanctuary Director and/or Shiwalik Conservator, and the Regional Committee would join in the selection process and approve the final candidate. This person must be innovative and wholly supportive of the community forest management model. The individual must have strong

and flexible interpersonal communication skills, and a technical forestry background.

Training Facilitator

This person would work on behalf of the Regional Committee to support and facilitate communication and training for the khol committees, minor forest produce committees, Range Panchayats, and the Sanctuary Committee. There would be one facilitator per range during the first three years in which the range comes under community management. The training facilitator role would be flexible, assisting the community management structure and the Regional Committee with the implementation process.

The facilitator's role would change over the three-year period. Initially the person would provide intensive education. The facilitator would assist all Van Gujjars and villagers in understanding the specific management structures and decision-making processes to be followed under the plan, focusing on encouraging positive communication within and between groups. This person would also assist individual deras in developing their proposals for the khol committee. The facilitator would be an additional resource person for the forest inventory, but he would not have decision-making powers within the community management structure. The facilitator is responsible for regular reporting to the RLEK Chairman, and the Regional Committee on the progress of his or her work.

Over the three-year term, the facilitator would gradually shift to a support role, and finally to an observer role during the third year. This shift in the role of the training facilitator in time is important, as it is a time limited training and facilitation position. In the long term, far less facilitation would be required, and it would be provided by the DFO position and the Regional Committee.

(a) Appointment: This person would be appointed and supervised by the RLEK Chairman and the Regional Committee would join in the selection process to approve the final candidate. This person must be innovative and wholly supportive of the community forest management model, and must also have strong and flexible interpersonal communication skills, and expertise in community development work.

Van Gujjars Choosing to Accept Agricultural Land Offers

Van Gujjars, who in the long term choose to accept a government offer of agricultural land, would be able to keep their buffalo in the khol for six months after moving, and return to the khol at any time during that period. After six months, if they choose to remain on the agricultural land, then their buffalo must vacate the khol. The khol committee could only sanction new buffalo in the khol if the khol is below its carrying capacity, and the buffalo are from areas which are above their carrying capacity.

Completion and/or Termination of the Agreement

At the end of the five year agreement for a range, the Regional Committee would review the successes and deficiencies in the plan, including the forest inventory. The agreement would automatically be renewed if:

- the forest inventory shows that the forest in the range has been maintained, or improved in quality, during the five year period, and
- the Range Panchayat requests a renewal of the agreement.

If the forest within the range deteriorates over the five year period, the Regional Committee would decide as to whether the agreement should be renewed.

The agreement could only be terminated within a range on the unanimous recommendation of all members of the Regional Committee.

Village Minor Forest Produce Committee Structure

Minor Forest Produce Committee Roles

The traditional rights and areas of use by each village, for each khol in the proposed park area to come under community management, would be defined by a consensus of the relevant villages and the Van Gujjars living in the area.

This has already been completed for all the khols in the year-one 1 pilot zone (see Chapter 11). The research required, to define boundaries in subsequent zones to come under community management, would be the responsibility of the evaluation staff person under the supervision of the Regional Committee.

Depending on the traditional rights and uses of each specific area, the village MFPC would be responsible for developing one or both of the following types of plans:

1. *Core Area Minor Forest Produce Management Plan:* This plan would define the management of the bhabbar grass resource in a particular khol, and provide a link with the khol committee with respect to bhabbar management. Liaison with the khol committee would be required about other minor forest produce including grazing or fuelwood collection, if there is any traditional practice of villagers entering the core managed by the Van Gujjars.

2. *Border Area Minor Forest Produce Management Plan:* This plan would define the management of a border area of the sanctuary, which is not under core area management, where the villagers have been the primary users and have traditionally collected minor forest produce.

In some areas the minor forest produce committee would have both these roles, while in some areas they would have one or the other. The responsibilities are described separately.

Minor Forest Produce Committee Membership

Every village man and woman who traditionally uses forest resources in the sanctuary would be eligible for membership in the Minor Forest Produce Committee (MFPC). A minimum of one-third of the members must be women, as they are important users of minor forest produce. Members of a particular MFPC would be able to use only the forest areas or resources in the proposed park defined for that MFPC. Non-members could not use forest resources in the proposed park. Elections would be held for an executive body of approximately twenty people for each MFPC. Executive members would serve a three-year term, but could be recalled sooner by a majority of the members of the minor forest produce committee at a full membership meeting. The village pradhan would be an ex-officio member of the MFPC executive. The executive would also have a minimum of one-third female members. Executive meetings would be open to all members of the MFPC. Two Van Gujjars from the relevant khol committee would participate in MFPC meetings, but would not have decision-making power.

Core Area Minor Forest Produce Management Plan

Committees, in villages which extract bhabbar grass from a specific khol, would be responsible for developing a minor forest produce management plan for this resource. The relevant committee would have all the resource in the area under its jurisdiction, and there would be no involvement of private contractors or Forest Corporation (Van Nigam) in the area. Any existing contracts with these bodies would be turned over to the MFPC for supervision at the outset of community management in the zone, and terminated at their expiry.

The full membership of the MFPC would meet to discuss and provide ideas for the development of the minor forest produce management plan. These ideas would be developed into a specific draft set of management procedures by the executive, which in turn would be submitted for approval by the consensus of the full membership. Finally, the MFPC plan would be submitted to the relevant Van Gujjar khol committee for its review and endorsement at a joint meeting. Unresolvable differences between the MFPC and the khol committee regarding the plan would be referred to the Range Panchayat for arbitration. The village panchayat pradhan and one other member of the MFPC would be nominated to participate in the relevant khol committee to maintain a regular liaison between the committees. The DFO would be available to assist if needed in the development and working of this plan.

The minor forest produce management plan would include the following:
* *Amount of Resource Extraction:* Definition of the amount of resource to be extracted throughout the year, the amount allotted to each household, and the procedures through

which each household would claim that allotment. There would be a documentation system to check when and how much of the resource each household was extracting (e.g., possibly a card or voucher system).

- *Extraction Procedures and Regulations:* Definition of appropriate methods and regulations for using the resource in a sustainable manner.

- *Planting and Conservation:* Specific plans to plant more bhabbar grass and improve the habitat.

- *Supervision System:* Procedures through which select members of the MFPC would supervise the above systems for utilising and managing the resource. This system would either involve rotating responsibility among members of the MFPC, or have the MFPC contribute to the salary of a one particular person.

Border Area Minor Forest Produce Management Plan

Village MFPCs, including some of those in the pilot zone, are the primary users of minor forest produce in border areas of the proposed park. In some instances Van Gujjars lop in these areas although they do not reside in them. The MFPC would be responsible for developing a border area management plan, regarding minor forest produce for their area according to the same process outlined above for the core area. It would include similar elements the core area plan:

- *Amount of Resource Use:* Definition of the amount of resource usage throughput the year (grazing and/or fuelwood collection), the amount allotted to each household, and procedures, if necessary through which each household would claim that allotment.

- *Extraction Procedures and Regulations:* Definition of appropriate methods and regulations for using the resource in a sustainable manner.

- *Planting and Conservation:* Specific plans to plant more trees and/or bhabbar grass to improve the habitat, including requests to the Sanctuary Committee for resources for improvement projects (i.e., tree seedlings or fencing materials).

- *Supervision System:* Procedures through which select members of the MFPC would supervise the above systems for utilising and managing the resource among members. This system would either involve rotating responsibility among members of the MFPC, or have the MFPC contribute to the salary of one particular person.

The traditional lopping rights of Van Gujjars in these areas would be recognised, and the relevant khol committee would submit their plans and procedures for lopping to the MFPC. These plans would be reviewed and endorsed by the MFPC, with disputes being referred to the Range Panchayat.

Revenue Status for Taungya Villages

All Taungya villages would be granted revenue status upon coming under the community forest management structure. Revenue status is necessary to give Taungyas security of tenure over their lands. This is essential to involve them in the long-term protection of the forests. They will take responsibility in the community management process, only if they can be assured some of the benefits. Revenue status would also give them access to rural development programs for the first time, and allow them to elect a gram sabha, which could be helpful in advocating community forest management policy.

Community Forest Protection Structure

Broad Protection Responsibilities

The essence of community forest management is that all members of the community area are responsible for forest protection. All Van Gujjars, and villagers extracting minor forest produce within a khol are responsible for the protection of the khol area, and must report violations to the khol committee or the Van Gujjar forest guards. The guards or committee must then report all illegal violations to the appropriate authorities, while the khol committee could take disciplinary action against anyone who is not abiding by khol or range practices. In these instances, khol committee decisions could be appealed to the Range Panchayat. Instances in which villagers are not abiding by the bhabbar extraction and management policies defined by the Villager Minor Forest Produce Committee would be referred to that committee for disciplinary action, through the villager responsible for supervision of the MFP plan, or the Van Gujjar forest guard. If the khol committee is dissatisfied with the village committee's action, they may appeal to the Range Panchayat to address the issue with the villager committee. Villagers responsible for supervising the extraction of minor forest produce for the MFPCs would be given authority, through the MFPC, to enforce MFPC regulations. Villagers would not be trained as full community forest guards at the outset as they are not forest dwellers, and have a more limited use of and investment in the forest. This policy could be reviewed by the Regional Committee in time if village MFPCs request training for individuals to guard boundary areas under their management.

Van Gujjar Forest Guards

Three Van Gujjars per khol would participate in a training process to become community forest guards with particular focus on protection of the local area. They would staff boundary checkpoints with Forest Department guards, and would also move across the interior area of khols for forest protection purposes. They would have full powers of forest protection equivalent to those of a Forest Department guard, but they would not have permission to carry weapons. Van Gujjar forest guards would receive remuneration on a monthly basis, at a level determined by the Range Panchayats, paid out of funds collected for Van Gujjar grazing and lopping fees. Bicycles would be made available for the guards' use. The khol committee would be responsible for supervising the Van Gujjar forest guards, and for ensuring that enough people remain in the khol in the summer for enforcement purposes. The guards would also play a role in assisting khol committees with forest improvement projects, e.g., bringing tree seedlings from the nursery.

(*a*) *Selection:* The Van Gujjars would be nominated by the khol committee for each khol, and the nominations would have to be approved by the Range Panchayat. Literacy is a pre-requisite.

Van Gujjar Forest Guard Training

All Van Gujjar forest guards would participate in a training course on the administrative and protection responsibilities of their role. The training would include a full review of the relevant laws and policies and would be defined and organised by the Regional Committee with particular responsibility to the Sanctuary Director, Shiwalik Conservator, RLEK Chairman, and the Advocate. Specific written materials would be provided in an understandable form for the participants. Forest Department forest guards, who are to work alongside Van Gujjar forest guards at the checkpoints, would participate in portions of the training as well.

Wireless Communication

RLEK would arrange for a wireless communication set to be provided for each khol committee, so that Van Gujjars can inform authorities rapidly if there are illegal activities that require backup from law enforcement authorities. RLEK would arrange for a unique frequency to be available for the community management structure to train in its use. The wireless may also be used for day to day communication, including requests for help in medical emergencies.

Forest Department and Police Enforcement Roles

Forest Department enforcement officials would staff boundary checkpoints along with Van Gujjar guards, but would not patrol or provide assistance within the khols, unless requested to do so by the khol committee, the village MFPCs, or the Van Gujjar forest guards. They would be expected to provide prompt back-up when requested, in line with their full enforcement powers. Police authorities would function in accordance with the same guidelines, in relation to their enforcement powers and roles. If the khol committee experiences problems in obtaining prompt support or effective co-operation from law enforcement authorities, they would lodge a complaint with the Regional Committee which would advocate on their behalf. Forest guards would maintain their role in patrolling closed areas where there is no khol committee.

The staffing level, for the Forest Department enforcement officials working in the community management area, would be determined by the Sanctuary Director and/or the Shiwalik Conservator in consultation with the Regional Committee. The level would be significantly lower than at present because of the additional Van Gujjar forest guards.

(a) Selection and Training: All Forest Department enforcement staff working in areas under community forest management would be appointed by the Sanctuary Director or the Shiwalik Conservator after discussion with the Regional Committee. Candidates would not be appointed to this role in a range if they are not acceptable to the Range Panchayat. Forest officials working in community management areas would participate in a training process arranged by the Regional Committee, including at least some aspects of the Van Gujjar forest guard training program.

Support for the Nomadic Lifestyle

Access to Forests giving for the Migration

The Chief Conservator of Forests for Uttar Pradesh would issue orders to Conservators and DFOs in the Hills Region, ordering officials to ensure free and unobstructed access to reserve forests for Van Gujjars during migration. The orders would demand prompt prosecution of people who threaten or harass Van Gujjars with respect to this use. Van Gujjar complaints would be reported to DFOs, and Van Gujjars would be asked to notify the Hill Monitoring Committee (see below) if problems continued.

Hill Monitoring Committees

The Chief Conservator of Forests for Uttar Pradesh would establish two Hill Monitoring Committees with respect to Van Gujjar presence in the Uttar Pradesh Hills, in the summer and monsoon seasons. One committee would serve the Uttarkashi area, and one the Tehri/Chamoli area. The Regional Committee would propose boundaries for committees, based on consultation with the Van Gujjars using the area and the Forest Department, in order to produce the most coherent geographical grouping of Van Gujjar summer areas. A Monitoring Committee could also be established for the area in Himachal Pradesh. Based on current research, Van Gujjars migrating to Himachal Pradesh did not report significant problems.

The Hill Monitoring Committee would include:
- The DFOs responsible for the area
- Four Van Gujjars from the area.
- RLEK representative
- Four village representatives from the area
- Other people as recommended by the existing committee

(a) Appointment and Tenure: The Van Gujjar nominations would be recommended to the Chief Conservator by a meeting of Van Gujjars who use the specific hill area. Recommendations for filling the villager positions would be made to the Chief Conservator only after the first meeting of the committee in the hills. They would be made by the new committee, based on the participants' knowledge of appropriate influential, and interested people in the area, who would be willing to serve. Appointments would be for two years, but could be renewed.

(b) Responsibilities: The committee would be responsible for hearing complaints from Van Gujjars, villagers, or forest officials with regard to conflicts during migration, or in the hills, between Van Gujjars and villagers or forest officials.

The committee could probably mediate in disputes, or suggest other path for resolution, but would not have independent authority. Only in part of the committee might review a situation, based on geographical proximity or jurisdiction. The committee would also be responsible for taking initiative in recommending guidelines for controversial aspects of Van Gujjar-villager interaction such as fees to be paid by Van Gujjars for passing through villages. The committee would determine a chairman within the first two meetings, and would meet at least three times per summer with a final meeting after the migration.

Support Programs for the Nomadic Movement

RLEK staff would consult Van Gujjars moving through each major migration path, in order to determine their needs so as to reduce problems on the migration. Support programs would be developed, based on Van Gujjar requests, within the philosophy of Van Gujjar self-help and responsibility. Programs might include water, fodder, and/or shelters at set locations. In addition, RLEK would consult Van Gujjars about the human and animal health team coverage needed in each area during the summer season.

Legal Education

Legal education would be included in the training for the khol committees to assist Van Gujjars in dealing with conflicts in the hills.

According to K. K. Sharma, Dy. Director (Senior), L. B. S. N. A. A., Mussoorie, 1995 "Section 35 of the Wildlife Act declaring certain areas as parks should be reviewed. The ground realities have to be kept in mind. Do our local conditions call for having Parks or Sanctuaries? Conservation is important, but it should not be at the cost of local dwellers. It is not possible to have parks in such areas. We may have sanctuaries, as in the latter we keep the rights of local dwellers in mind. One of the basic aims of any Act is to see whether it can be implemented or not. Any Act which is far away from reality only encourages corruption especially at the lower levels. The transhumance pastoralists should be encouraged to follow their traditional practice. Their cattle provide manure on the route, encourage cross-fertilisation of flora and fauna, and also provide a source of pure milk to the surroundings."

Van Gujjar Development Priorities

Fee Collection and Uses

An essential component of community management is control of, and accountability for, financial resources. Lopping and grazing fees therefore would be paid to the Range Panchayat, rather than the Forest Department, under the community forest management structure. The Range Panchayat would set the rate after consultation with the khol committees. Van Gujjar payments would be considerably less under the new system, even if official rates are increased, as there would be no unofficial fees. The fees would be paid to a designated Van Gujjar financial officer for the range, who would operate a bank account under the supervision of the Range Panchayat. The range financial officer would initially be trained and monitored by an Accountant working in the training component. The officer would provide monthly financial statements of the money received and spent for the Range Panchayat, the Sanctuary Committee, the Regional Committee and the khol committees. The accounts would be audited yearly. All funds paid by deras in a khol would be returned to that khol, through funding of the forest guards and programs proposed by the khol's Health and Education Committee.

Health and Education Committee

The Khol Committee would appoint a Health and Education Committee responsible for defining health and education programs to be provided in the khol. This committee would have the same number of men and women for a balance of male and female priorities in defining programs. The committee would fund and supervise a teacher to provide education to the deras in the khol. Any additional funds would be available for programs as the committee sees fit. Programs would be flexible to accommodate the needs and schedules of those going to the hills. The committee would report to the khol committee regularly on the basis of its work. Meetings would be open to everybody in the khol.

Human and Veterinary Health Programs

The Health and Education committee would work with the Van Gujjar para-medic in the area, and representatives of the health team from RLEK, to define what is needed from the health team. Emphasis would be placed on preventive health programs, including family planning, vaccinations, nutrition awareness, and drinking water safety. Health programs would be provided in the hills during summer and monsoon seasons.

Khol Forest Improvement Projects and the Tree Nursery

Khol Committees could propose and carry forest improvement or habitat preservation projects, such as slowing soil erosion, protecting fragile areas, planting of vegetation to meet specific wildlife needs,. Funds could be obtained from the collection of Van Gujjar fees, or from budget through the Sanctuary Committee and Regional Committees. The DFO might assist or facilitate such a project, but all projects in the khol would require the support and approval of the khol committee. If paid labourers are required for the project, the khol committee would determine if they are to be Van Gujjars or hired from outside.

Funds allocated to projects would be dispensed through the range bank account, with regular statements to account for expenditures. The khol committee would be accountable to the Regional Committee for the expenditure of funds, and the successful completion of work which is under khol committee supervision.

A tree nursey would be operated for each sanctuary with staff initially supported by the training component.

Economic Options Programming

RLEK staff would work with khol committees to strengthen the milk marketing co-operative. Requests from Van Gujjars for the development of other economic options would be pursued, e.g., options for traditional craft embroidery by Van Gujjar women, tourism etc.

Public Education, Tourism and Interpretation

In the initial phases of the community management structure, Public Education would be through the work of the Regional Committee, and the activities of the staff person responsible for evaluation and documentation of the progress of the community management structure. This would involve the release of written reports, and public speaking by members of the Regional Committee, Van Gujjars, and villagers, about the purpose, strategies, and accomplishments of the new management structure. It is important to acquaint the public with the progress of this new approach to the management of protected areas.

Tourism and interpretation activities are important in the long-term development of the park, as a means to increase environmental awareness in the surrounding region and promote protection of the forest area. Tourism and interpretation activities however, would not be a priority in the initial phases of the community management structure in the pilot area (see below). It would be important to reduce the number of complicating factors, while restructuring the management system. The pilot area has not previously been open to the public, who could have a major impact on wildlife and/or local people. Open access could also cause poachers and tree fellers to justify their presence in the area. Finally, Van Gujjars and local villagers have not had the exposure, and opportunity to discuss their perspectives and roles in tourism and education, a step which must occur to insure that tourism and public education are compatible with the community management philosophy.

Proposals for tourism and interpretation activities, in later phases of the community management process, would be developed by the Regional Committee in consultation with range panchayats, khol committees, and forest officials.

Training Component

A major training component is essential to facilitate the major shift to community management. The total proposed area for community management is divided into four zones, with one new zone added to community management each year (see Chapter 11). The training component in each zone would cover three years, thus, it would take six years to complete the training phase in all four zones.

Table 16
Schedule and Duration of the Training Component

Phase	No of Ranges	#1	#2	#3	#4	#5	#6
Year 1 Zone	2	X	X	X			
Year 2 Zone	3		X	X	X		
Year 3 Zone	2			X	X	X	
Year 4 Zone	2				X	X	X

The overall amount of training done would vary through the years, with the lowest levels in the first and last years when only one zone is being trained, and the highest levels in the middle years, when three zones would be involved in training simultaneously. The staffing and infrastructure requirements for each zone are defined below. There are three parts to the training component: (1) training facilitators, (2) training for the community management committees, and (3) training for the community forest guards.

The training component would be conducted by the training facilitator for each range working under the direction of the Regional Committee. The training would be planned in close consultation with the relevant committees at each level, to respond to their needs and requests. The training would draw on resource people from organisations represented on the Regional Committee, and from a broader range of resource people and organisations as required.

Training Facilitator

This person would work on behalf of the Regional Committee to facilitate communication and training for the Khol Committees, Minor Forest Produce Committees, Range Panchayats, and the Sanctuary Committee.

Training for the Community Management Committees

(a) *Khol Committee and Minor Forest Produce Committee Training:* Each Khol Committee and Minor Forest Produce Committee would require training in order to take over responsibility for management of their areas. This process must continue over a three-year period, as new problems and issues will develop through time. It is essential to have sufficient support to help the committees take on their new responsibilities.

Training would be at an individual committee level, although the Khol Committee and minor forest produce committee, who have common resource issues, would be combined for parts of the training. There would also be participation from forest officials, working under the community management system in the training as appropriate. The training schedule and outline of broad topics for each committee is listed below (see Table 17). The detailed curriculum would be defined by the Regional Committee through the training facilitators and would respond to committee requests and needs. The training would not be restricted only to formal sessions, but also informal and continuous through regular interactions between the training facilitator and the members of the committees. The curriculum during the training sessions would involve the committee tackling some of its management issues and responsibilities.

(b) Training for Range Panchayat and Sanctuary Committee Members: These Range Panchayat and Sanctuary Committee members for each zone would also participate in twelve days of training throughut the three-year period. The training would follow the same times and schedule as outlined above for the khol and minor forest produce committees. The members of these committees for each zone would be trained as one group. The topics would follow the pattern outlined for the Khol Committees and MFPCs except that the perspective and detail on each topic would change in relation to the differing roles, responsibilities and needs of the range panchayats and sanctuary committee. There would be training in the administration of financial accounts.

(c) Training for Regional Committee: The Regional Committee would identify its own training needs, and rely on the training facilitators in the year-one zone to make arrangements as appropriate. In particular, there would be more emphasis placed on research and evaluation, legal procedures, administrative committee work, and staff supervision roles.

(d) Training of Range Financial Officers: The range financial officers would receive detailed training and supervision in development, administration, and reporting the range bank accounts.

Training for Community Forest Guards

All community forest guards would be trained as a group for each phase of the implementation of community management. There would be 26 days of training for each group throughout their three year training period. The schedule, spacing, and outline of topics is described below. A detailed curriculum would be defined by the Regional Committee.

Table 17
Training for Khol Committees and Minor Forest Produce Committees

Time Frame	Topics
3 days at start	Detailed review and signing of agreements
	Laws, rights, responsibilities, and decision-making
	Conflict resolution and mediation strategies
	Forest carrying capacity concepts, and dera and MFP plans
	Problem identification and resolution strategies
	Financial, administrative, and reporting procedures
	Relationship and roles vis a vis Forest Department
	Development for individual dera plans
	Relationships and powers, between MFPC & khol committees
	Supervision and roles vis a vis community forest guards
	Legal issues in conflicts in the hills
2 days at 3 months	Review of dera and/or minor forest produce plans
	Training for forest inventory
	Review of management procedures and outstanding issues
	Joint discussions between khol and minor forest produce committees
1 day at six months	Review of outstanding legal and enforcement issues
	Review of community forest guard roles
2 days at 12 months	Review of administrative issues
	Review of community forest guard roles
	Joint discussions between khol and minor forest produce committees
1 day at 18,24,30,&36 months	Topics to be defined in relation to committee needs and requests

Table 18:
Training for Community Forest Guards

Time Frame	Topics
10 days at start	Detailed review of job description and jurisdiction Detailed review of community management structure Review of legal processes and procedures Indian Forest Act of 1927, Wildlife Protection Act 1972 Protection procedures and documentation Reporting and accountability structures Use and maintenance of wireless Relationship and powers vis a vis forest officials Relationship to khol committees and MFPCs Communication skills for forest guard role Problem solving skills in relation to forest guard role Case studies and scenarios Discussion with Forest Department Forest Guards
3 days at 3 months	Case studies and scenarios Problem-solving and communication skills Review of legal processes, procedures and laws Review of enforcement procedures and implementation problems
2 days at 6, 12, 18, 24, 30, & 36 months	Topics to be defined in relation to needs and requests

Financial Costs and Organisational Commitments

On-Going Costs

The community management structure as proposed would reduce current government expenditure in relation to the present system, in that there would be a lower number of forest officials required, with the specific numbers to be determined by the Sanctuary Director and Shiwalik Conservator in consultation with the Regional Committee. The only new government funded positions would be that of the DFO for each sanctuary. Exact expenditures would be calculated, based on a budget to be developed by the Regional

Committee in consultation with the Sanctuary Director and the Shiwalik Conservator. Existing routine Forest Department funds for forest improvement of the area should be transferred to the Regional Committee for the community management area, as part of the budgeting process. The Forest Department would continue to be responsible for maintaining the sanctuary roads and buildings, according to existing budgets.

Initial Training and Infrastructure Costs

The training component and initial infrastructure would entail a significant initial expenditure but would not require continual support given its time limited function. This expenditure would be funded through external support arranged through the NGO (RLEK). The cost would involve the following :

1   evaluation and documentation staff position assigned to the Regional Committee
1   accountant to train range financial officers (part-time)
1   training facilitator per range (3 year term position)
12 days evaluation etc. training for each khol committee
12 days accountant training for each minor forest produce committee
12 days training facilitation etc. for Range Panchayat and Sanctuary Committee members
1   wireless per khol committee
3   bicycles per khol committee (for community forest guards)
6   regular staff for tree nurseries
    casual staff for tree nurseries
    supplies for tree nurseries

Administration of Funding for the Training Component

Funding for the training component would be administered by RLEK, and given to the Regional Committee for spending expenditure as it is a multi-agency committee but without legal autonomy as an independent organisation. In the same manner, the Forest Department would administer the funds it gives to the Regional Committee for the DFO positions, and the routine funds for forest improvement.

234

Chapter 11
Area Specifications and Pilot Implementation

Areas and Phases of Implementation

It is important to introduce a management change of this size gradually, to identify strengths and learn from problems as the process evolves. This proposal sets out detailed specifications for implementation in the year 1 pilot zone, and then recommends implementation in three additional areas on a sequential yearly basis to bring all areas of the proposed Rajaji Park under community management (see Map 8). The evaluation member of staff, working under the supervision of the Regional Committee, would complete the additional detailed research on boundaries for khol committees and village MFPCs in each new zone. The implementation schedule is proposed as follows :-

Year 1 Zone: Chillawali Range and Mohand Areas
Year 2 Zone: Ramgarh, Dholkhand, and Shakumber Block
Year 3 Zone: Ranipur, Kansrao, and Motichur Ranges
Year 4 Zone: Chilla and Gohri Ranges

The following table defines the number of ranges, khol committees, and MFPCs for each phase. The numbers and boundaries of khol committees and MFPCs have been precisely defined for the year 1 pilot phase. The specific boundaries and numbers would be defined by the Regional Committee for years 2-4. The numbers are estimated however, for these subsequent years, based on the distribution of Van Gujjars and villages in them.

Table 19
Specifications for Phases

Phase	No. of Ranges	No. of Khol Committees	No. of MFPCs
Year 1	2	13	8
Year 2	3	13	10
Year 3	2	10	9
Year 4	2	10	10

Year 1 Pilot Zone

The Chillawali and Mohand Areas (Mohand to Sahasra Blocks) have been chosen for the pilot zone for year 1, in that they represent accessible zone for monitoring purposes. Mohand is already a communications centre for Van Gujjars, and a base of operations for the Forest Department and RLEK. Van Gujjars in this area are particularly keen to be the first to implement community management, and there is also support and commitment from the villagers who use the zone. This area cuts across two Forest Department jurisdictions, the proposed park area and the Shiwalik Forest Division. Different management issues would be likely to arise in each area, given the differences in law and management practice. Thus, maximum learning would result in this first pilot phase. The Shakumber block in the Mohand area is left until year two as there are more complex relationships with villagers due to the presence of a temple. Lopping is also prohibited in Shakumber at present.

Pilot Zone Specifications

maps 9 and 10 provide a detailed view of the pilot area including charts on range specifications. There are 1249 Van Gujjars and 2588 buffalo in the Chillawali Range, and 1536 Van Gujjars and 3181 buffalo in the Mohand area (see Table 3). Chillawali would be divided into four khol committees and Mohand into five khol committees (see Maps 9 &10). There are two Forest Department checkpoints to be staffed in Chillawali, and three in Mohand.

Elephant Corridor Proposals

The elephant corridor falls in a zone of border area villager use, which comes under community management in year 3 and therefore, detailed boundary proposals for this area would be defined by the evaluation staff member and the Regional Committee. The opening of this corridor however, is of concern to all, and it is recommended that the park authorities proceed immediately with clearing the area by negotiating relocation of the army munitions dump, negotiating a just settlement for the re-location of the Tehri Dam evacuees, and proceeding with the construction of a bridge to help elephants crossing the Chilla Power Canal. The army munitions dump must be moved first as that may be the most difficult negotiation process, and it would be useless and unjust to move the Tehri Dam evacuees, only to have the corridor still blocked by the munitions dump. The Regional Committee would advocate for these proposals to the government.

It is essential to begin the above steps in year 1, so that the area will be cleared for community management in year 3. A villager minor forest produce committee to manage this border area is essential. Sources of specific elephant fodder would have to be planted in the corridor, if the elephants are to be encouraged to use it. Villagers however, are already using this area for grazing, and so there would be competition between elephants and villager cattle in this narrow zone. Arbitrary restrictions by the park authorities on grazing in the corridor is likely to fail as in other areas and so it is essential for park authorities to work out management agreements for the corridor with the villagers who use it.

CONCLUSION

This report defining Van Gujjar proposals for Community Forest Management in Protected Areas (CFM-PA), is the first step to defining a new approach to protected area management which protects both the ecosystem and the indigenous people living within it. Community management must necessarily proceed on a case by case basis as structures and approaches will vary depending on the characteristics and needs of the local communities and the environment. Community Forest Management must also be an evolving process, as initial proposals will develop as local people work together with resource staff and organisations. The Van Gujjars in the Rajaji context provide an excellent test case through which to begin to analysing the strengths and areas for development of the model over a period of time.

The specific strategies proposed in this report address the major problems of the Rajaji Park area over the coming years:

The *Core Area Community Forest Management Structure* would give Van Gujjars the sense of security and the incentives they require in order to take responsibility for improving the forest in which they have been living for centuries. The process of the khol committees would provide a peer community context which would cause all Van Gujjars to monitor their own impact on the forest. The Range Panchayat and the Sanctuary Committee would provide an effective co-ordination and conflict resolution mechanism, to balance the needs of Van Gujjars and the ecosystem. The Regional Committee would ensure a public, accountable process, and a means to approach the government on behalf of community-based forest policy.

The *Villager Minor Forest Produce Committee Structure* would give villagers their secure, traditional rights to use of bhabbar grass, whilst for the first time giving them responsibility for maintaining and improving the tracts in the khol areas. In addition, by meeting their needs, the villagers would be more open to assisting in the protection of the forests in consultation with Van Gujjars. They would also take

239

responsibility for the management of border areas where they have been the primary users. The community management structure provides effective links between Van Gujjars and villagers so that they could work together co-operatively for the betterment of all and the ecosystem.

The *Community Forest Protection Structure* would utilise the eyes and ears of every Van Gujjar living within the area, and villagers using the area, as an asset in the protection of the forest against illegal activities. It would provide trained, reliable, and honest individuals in protection roles as Van Gujjar forest guards. They would have a wireless communication system and back-up support from law enforcement authorities, with the Regional Committee providing a means to press the law enforcement officials to follow through in cases when there are difficulties.

The measures in *Support for the Nomadic Lifestyle* begin to address the objective to Van Gujjars' moving to the hills in the summer season through Forest Department orders giving them access to reserve forests on the migration, and Hill Committees to assist in resolving conflicts with villagers. In addition, the change from the present management system to the Community Forest Management structure would remove the pressures Van Gujjars now feel to stay in the park area in the summers to protect their deras and forests.

Finally, the focus on *Van Gujjar Development Priorities* would simultaneously increase Van Gujjars' sense of security with respect to their lifestyle and culture, and provide them with options in the future for people choosing to leave the forest. Education, health, and family planning programs would play a positive role in lowering infant mortality, and in allowing parents to choose family size which in turn would make the forest lifestyle more sustainable over the long term.

These five major strategies complement each other in protecting the ecosystem; hence the whole is greater than the sum of its parts. Forest deterioration would be addressed through tree planting and stopped effective enforcement against tree fellers. The depletion in ground cover would be through weeding, adjusting buffalo herds to the carrying capacity of the area, and giving the forest more rest through nomadism. Wildlife would be supported and protected through measures effective against poachers, the planting of fodder trees and the improvement of forest quality and ground vegetation. The impact of fires would be reduced with the mobilisation of Van Gujjars to maintain firelines and provide a prompt response in emergencies.

Finally, community forest management would bring a more harmonious and positive interpersonal environment for Van Gujjars, villagers, and forest officials living and working in the Rajaji area. It would allow all of them to participate in a system in which incentives and values foster forest protection and personal respect.

REFERENCES

Agarwal, A., & Narain, S. (1992). *Towards a Green World: Should Global Environmental Management be Built on Legal Conventions or Human Rights?* Delhi: Centre for Science and Environment.

Agarwal, A., & Narain, S. (1993). People-Oriented Schemes Can Save Sanctuaries. *The Economic Times*, June 20.

Ahmed, A.S., & Hart, D.S. (1984). *Islam in Tribal Societies*. New Delhi: Routledge.

Ahuja, C. (1994). Elephant Population Soars. *Indian Express*, February 17.

Archibold, G., & Davey, S. (1993). Kuna Yala: Protecting the San Blas of Panama. In E. Kemf (Ed.) *Protecting Indigenous Peoples in Protected Areas: The Law of the Mother*. San Francisco: Sierra Club Books, 52-57.

Bahuguna, V.K. (1993). *Collective Resource Management: An Experience in the Harda Forest Division*. Bhopal: Regional Centre for Wastelands Development.

Berkmuller, K. (1986). *Pressure and Dependency by Local Peoples on the Resources of Rajaji National Park*. Dehra Dun: Wildlife Institute of India Research Project.

Bhatia, A.K. (1995). *A Study on Rajaji National Park and Socio-Economic Life of Local Inhabitants (Van Gujjars)*. Dissertation Report, Dept. of Rural Development, Xavier Inst. of Social Service, Ranchi.

Bhatnagar, Y.V. (1991). *Habitat Preference of Sambar (Cervus unicolor) in Rajaji National Park*. M.Sc. Dissertation in Wildlife Science submitted to Saurshtra University, supervised by the Wildlife Institute of India, Dehra Dun.

Bhatt, S.D. (1993). *Habitat Use by Chital (Cervus Axis) in Dhaulkland, Rajaji National Park, India*. M.Sc. Dissertation in Wildlife Science submitted to Saurshtra University, supervised by the Wildlife Institute of India, Dehra Dun.

Bhatt, J. (1993b). Need to Preserve Eco-Systems. *Times of India*, February 25.

Bhattacharyya, A. (1995). Commercial Ventures Gnawing through Wildlife Bastions. *The Pioneer*, New Delhi, June 24.

Binghay, A.H. (1899). *History, Castes and Culture of Jats and Gujjars*.

Centeno, J.C., & Elliot, C. (1993). Forest Home: The Place Where One Belongs: Yanomami of Venezuela. In E. Kemf (Ed.) *Protecting Indigenous Peoples in Protected Areas: The Law of the Mother.* San Francisco: Sierra Club Books, 95-103.

Centre for Science and Environment (1985). *The Second Citizens' Report on the State of India's Environment.* Delhi: Centre for Science and Environment.

Chalawadi, S.M. (1995). JFM- Extension to Wildlife Sanctuaries in Karnataka. In S.B. Roy (Ed.), *Enabling Environment for Joint Forest Management.* New Delhi: Inter-India Publications, 139-144.

Chandra, N.S., & Poffenberger, M. (1989). Community Forest Management in West Bengal: FPC Case Studies. In K.C. Malhotra (Ed.) *Forest Regeneration through Community Protection: The West Bengal Experience.* West Bengal: West Bengal Forest Department, 22-47.

Cincotta, R., & Pangare, G. (1993). Pastoralists: Brokers of Agricultural Soil Fertility. *Wasteland News, VIII (3),* 40-44.

Clark, A., Sevill, H., & Watts, R. (1986) *Habitat Utilization By Gujjar Pastoralists in Rajaji Wildlife Sanctuary.* Dehra Dun: Wildlife Institute of India.

Cordell, J. (1993). Boundaries and Bloodlines: Tenure of Indigenous Homelands and Protected Areas. In E. Kemf (Ed.) *Protecting Indigenous Peoples in Protected Areas: The Law of the Mother.* San Francisco: Sierra Club Books, 61-68.

Cornista, L.B., & Escueta, E.F. (1989). Communal Forest Leases as a Tenurial Option in the Philippine Uplands. In M.Poffenberger (Ed.), *Keepers of the Forest:Land Management Alternatives in Southeast Asia.*

Dasman, R. (1988). National Parks, Nature Conservation, and "Future Primitive". In Bodley (Ed.), *Indigenous Peoples & Development Issues,* Mayfield California, 301-309.

Deb Roy, R.K., Shankarnarayan, K.A., & Pathak, P.S. (1980). Fodder Trees of India and Their Importance. *Indian Forester, 106,* 306-311.

Down to Earth (1995). Forests for Whom? *Down to Earth, 4 (15),* December 31, 49-55.

Dupree, L. (1980). *Afghanistan.* Princeton, NJ: Princeton University Press.

Dwivedi, A.P. (1992). *Principles and Practices of Indian Silviculture.* Dehra Dun: Surya Publishers, Chap. 6.

Edgaonkar, A. (1995). *Utilisation of Major Fodder Tree Species with Respect to the Food Habits of Domestic Buffaloes in Rajaji National Park, India.* M.Sc. Dissertation in Wildlife Science submitted to Saurshtra University, supervised by the Wildlife Institute of India, Dehra Dun.

Gooch, P. (1988). *Nomadic Muslim Gujjars, A Pastoral Tribe in North India. A Case for Survival.* Dehra Dun, UP: RLEK.

Gooch, P. (1992). Transhumance Pastoralism in Northern India: The Gujjar Case. *Nomadic Peoples, 30,* 84-86.

Gooch, P. (1994). *Conservation for Whom? Van Gujjars and the Rajaji National Park in India.* Unpublished document, RLEK, Dehra Dun.

Gooch, P. (1995). When the Land Shouts. *Down to Earth, 3 (18),* February 15, 46-47

Goudie, A. (19). *The Human Impact on the Natural Environment.* Oxford, UK: Blackwell, 34-37.

Gould, J., & Lees, A. (1993). God of the Forest Forever. In E. Kemf (Ed.) *Protecting Indigenous Peoples in Protected Areas: The Law of the Mother.* San Francisco: Sierra Club Books, 77-80.

Government of India (1976). *Report of the National Commission on Agriculture, Part IX, Forestry.* New Delhi: Department of Agriculture and Cooperation, Government of India.

Gray, A. (1991). *Between the Spice of Life and the Melting Pot: Biodiversity Conservation and Its Impact on Indigenous Peoples.* Copenhagen: International Working Group on Indigenous Affairs (IWGIA).

Gray, A. (1992). The Impact of Biodiversity Conservation on Indigenous Peoples. In V. Shiva (Ed.), *Biodiversity: Social and Ecological Perspectives,* Dehra Dun: Natraj Publishers, 59-76.

Gupta, J.R. (1995). Evolution of a Legal Agreement for Joint Forest Management in Harayana. *Wasteland News, XI(1),* 38-48.

Hiremath, S.R., Kanwalli, S., & Kulkarni, S. (1995). Forestry as if People Mattered. In S.R. Hiremath, S. Kanwalli, S. Kulkarni (Eds.), *All About Draft Forest Bill and Forest Lands,* Dharwad: Samaj Parivartan Samudaya.

ILO (1989). *Convention Concerning Indigenous and Tribal Peoples in Independent Countries.* Copenhagen: International Labour Organization (ILO).

Indian Institute of Public Administration (1994). *Biodiversity Conservation Through Ecodevelopment, An Indicative Plan.* Prepared on behalf of the Government of India, Ministry of Environment and Forests and the concerned state governments for submission to the United Nations Development Programme. Delhi: Indian Institute of Public Administration.

International Alliance of the Indigenous-Tribal Peoples of the Tropical Forests (1993). Charter of the Indigenous-Tribal Peoples of the Tropical Forests. In W. Fernandes, *The Indigenous Question: Search for an Identity,* New Delhi: India Social Institute, 181-187.

Iqbal, S.M., & Nirash, K.L. (1978). *The Culture of Kashmir.* Marwah Publishing, see section by L.R. Publisher.

Jena, N.R. (1993). *National Parks and Sanctuaries in India: Some Issues of Concern.* Paper presented for the National workshop on "Declining Access to and Control over Natural Resources in National Parks and Sanctuaries," October 28-31, Dehra Dun. Society for Participatory Research in Asia and Rural Litigation Entitlement Kendra.

John, B.K. (1995). The Last Post for the Tiger and the Elephant. *The Pioneer,* July 16.

Johnsingh, A.J.T., Narendra Prasad, S., & Goyal, S.P. (1990). Conservation Status of the Chilla-Motichur Corridor for Elephant Movement in Rajaji-Corbett National Parks Area, India. *Biol Conserve, 51,* 125-138.

Johnson, N. (1994). Public and Community Participation in the Global Environment Facility's Funding of the Convention on Biodiversity. In A.F. Krattinger, J.A. McNeely, W.H. Lesser, K.R. Miler, Y. St. Hill, and R. Senanayake (Eds.), *Widening Perspectives on Biodiversity (Asian Edition),* Dehra Dun: Natraj Publishers, 401-408.

Johnson, S. (1995). Clarified JFM Concepts. In S.B. Roy (Ed.), *Enabling Environment for Joint Forest Management,* New Delhi: Inter-India Publications, 47-58.

Joshi, D. (1995). Massive Protest Greets Tree Felling. *Indian Express*, September 30, p. 6.

Kaushal, A. (1993). Teaching Nomadic Women in the Wilds. Dehra Dun: Rural Litigation Entitlement Kendra.

Kemf, E. (1993). *Protecting Indigenous Peoples in Protected Areas: The Law of the Mother*. San Francisco: Sierra Club Books.

Khanna, D.P.S. (1992). *Glimpses of Indian Tribal Life*. Delhi:Sarita Book House, Ch. 29.

Khatana, R.P. (1992). *Tribal Migration in Himalayan Frontiers*. Gurgaon, Haryana: Vintage Books.

Kothari, A. (1994). People's Participation in the Conservation of Biodiversity in India. In A.F. Krattinger, J.A. McNeely, W.H. Lesser, K.R. Miler, Y. St. Hill, and R. Senanayake (Eds.), *Widening Perspectives on Biodiversity (Asian Edition)*, Dehra Dun: Natraj Publishers, 137-146.

Kothari, A. (1995). *Conserving Life: Implications of the Biodiversity Convention for India*. New Delhi: Kalpavriksh.

Kothari, A. (1995b). *JPAM Update*. November, New Delhi: Indian Institute of Public Administration.

Kothari, A., Bande, S., Singh, S., & Variava, D. (1989). *Management of National Parks and Sanctuaries in India: A Status Report*. New Delhi: Indian Institute of Public Administration.

Kothari, A., Suri, S., & Singh, N. (1995). Conservation in India: A New Direction. *Economic and Political Weekly,* October 28, 2755-2766.

Krattinger, A.F., McNeely, J.A., Lesser, W.H., Miller, K.R., St. Hill, Y., and Senanayake R. (Eds.) (1994). *Widening Perspectives on Biodiversity (Asian Edition)*. Dehra Dun: Natraj Publishers.

Kumar, D. (1995). *Management Plan of Rajaji National Park, Vols. I & II*. Dehra Dun, Uttar Pradesh: Rajaji National Park.

Lidhoo, M.L. (1988). *Kashmir Tribals*. Srinagar: Minakshi Publishers.

Lohman, L. (1992). Who Defends Biodiversity? Conservation Strategies and the Case of Thailand. In V. Shiva (Ed.), *Biodiversity: Social and Ecological Perspectives*, Dehra Dun: Natraj Publishers, 83-102.

Mandosir, S. & Stark, M. (1993). Butterfly Ranching. In E. Kemf (Ed.), *Protecting Indigenous Peoples in Protected Areas: The Law of the Mother*. San Francisco: Sierra Club Books, 114-120.

McNeely, J.A. (1992). *The Contribution of Protected Areas to Sustaining Society*. Opening address to the *IVth World Congress*, UNEP and UNDP, Caracas, Venezuela, Feb. 10-12.

McNeely, J.A. (1993). People and Protected Areas: Partners in Prosperity. In E. Kemf (Ed.) *Protecting Indigenous Peoples in Protected Areas: The Law of the Mother*. San Francisco: Sierra Club Books, 249-257.

McNeely, J.A., Miller, K., Reid, W., Mittermeier, R., & Werner, T. (1990). *Conserving the World's Biodiversity*. IUCN, Gland, Switzerland; WRI, CI, WWF-US, and the World Bank, Washington, D.C.

Ministry of Environment and Forests (1988). *National Forest Policy, 1988*. No. 3-1/86-FP. New Delhi: Ministry of Environment and Forests, Government of India.

Mitra, A. (1993). Chipko: An Unfinished Mission. *Down to Earth*, April 30, 25-36.

Mukhia, S. (1994a). Saving to Survive. *Down to Earth*, November 15, 22-24.

Mukhia, S. (1994b). The Roots of Prosperity. *Down to Earth*, December 15, 22-23.

Munshi, K.M. (1955). *The Glory that was Gujjar*. Bombay: Desh.

Nair, S.C. (1993). *Biosphere Reserves in the Indian Context*. New Delhi: Indian National Trust for Art and Cultural Heritage.

Narain, S. (1995). Debate on Management of Protected Areas, October 10, 1995, Centre for Science and the Environment, Delhi. Reported in *Down to Earth, 4(15)*, p. 53.

Natraj Publishers (1992). *The Wildlife Protection Act, 1972*. Dehra Dun: Natraj Publishers.

Palit, S. (1992). Implementation of Joint Forest Management. *Wasteland News, VIII(2)*, 58-66.

Poffenberger, M., & Sarin, M. (1995). Fibre Grass from Forest Land. *Society and Natural Resources, vol. 8*, 219-230.

Pottie, P.S. (1995). *Preliminary Report on the Rajaji National Park.* Bombay: The Indian People's Tribunal on Environment and Human Rights.

Prasad, R., & Bhatnagar, P. (1988). *Grazing Forests: An Ecological Liability.* Jabalpur: State Forest Institute.

Raju, G. (1995). Forest Protection Committees: Some Institutional Aspects. In S.B. Roy (ed.), *Experiences from Participatory Forest Management,* Delhi: Inter-India Publications, 91-103.

Ral, U. (1995). Tigers on the Run in MP Reserves. *Indian Express,* July 7.

Ramesh, M.K. (1996). Wildlife Protection Act, 1972: An Agenda for Reform. Paper presented at the National Consultation Workshop on Community Forest Management in Protected Areas, sponsored by RLEK, Dehra Dun, February 24-25.

Ramprakash, M.L. (1995). JFM in National Parks and Sanctuaries. In S.B. Roy (Ed.), *Enabling Environment for Joint Forest Management.* New Delhi: Inter-India Publications, 133-138.

Rathore, B.M.S., & Campbell, J. (1995). Evolving Forest Management Systems: Innovating with Planning and Silviculture. *Wasteland News, IX (1),* 4-19.

Raven, P. (1988). Our Diminishing Tropical Forests. In E.O. Wilson (Ed.), *Biodiversity.* Washington: National Academy Press.

Rawat, J.K. (1993). Economic Considerations in Tree Lopping (Acacia Nilotica). *Indian Forester, 119,* 703-707.

Rawat, A. S. (1993b). The Gujjars of Tarai and Bhabhar. In A.S. Rawat (Ed.), *Man and Forests; The Khatta and Gujjar Settlements of Sub-Himalayan Tarai.* New Delhi: Indus Publ. Co.

RLEK (1994). *RLEK.* Information pamphlet. Dehra Dun: Rural Litigation Entitlement Kendra.

RLEK (1995). *What They Say.* Dehra Dun: Rural Litigation Entitlement Kendra.

RLEK (unpublished). Rate of compensation for Death/Disablement of Human Beings and Cattle Caused by Wildlife. File document.

Sale, J.B., Chaudhury, S., & Khan, A. (1989). Ranging and Feeding Patterns of A Rajaji Tusker. In E.G. Silas, M.K. Nair, and G. Nirmalan (Eds.), *The Asian Elephant: Ecology, Biology, Diseases,*

Conservation, and Management. Trichur: Kerala Agricultural University.

Sarabhai, K.V., Bhatt, S., & Raju, G. (1991). *People's Involvement in Wildlife Management: An Approach to Joint Sanctuary Management of the Shoolpaneshwar Wildlife Sanctuary, Gujarat.* Ahmedabad: VIKSAT.

Sarin, M. (1993). JFM in Harayana. *Wasteland News, VIII(4),* 3-4.

Sarin, M. (1995). Community Forest Management: Where are the Women? In *The Hindu Survey of the Environment 1995,* Madras: S. Rangarajan, 25-29.

Sarkar, S.K. (1995). Joint Forest Management in West Bengal at the Crossroads. In S.B. Roy (Ed.), *Experiences from Participatory Forest Management,* Delhi: Inter-India Publications, 104-110.

Saxena, N.C. (1995). *Forest, People, and Profit: New Equations for Sustainability.* Dehra Dun: Natraj Publishers.

Saxena, N.C. (1995b). *Toward Sustainable Forestry in the U.P. Hills.* Mussoorie: Lal Bahadur Shastri National Academy of Administration.

Sengupta, S. (1993). A Few Unresolved Issues in Harayana. *Wasteland News, VIII(4),* 40-45.

Shah, S.A. (1993). Important Imperatives of Joint Forest Management. *Wasteland News, VIII(4),* 38-39.

Shah, S.A. (1995). Status of Indian Forestry-I. *Wasteland News, IX(2),* 14-32.

Sharma, K.K. (1995). Parks, Policy and People: A Case Study of Rajaji National Park. *The Administrator, Vol. XL,* 183-200, June.

Sharma, S.K., & Gupta, R.K. (1981). Effect of Seasonal Lopping on the Top Feed (Loong) Production and Growth of Prosopis Cineraria (Linn.) Druce. *Indian Journal of Forestry, 4 (4),* 253-255.

Shashi, S.S. (1979). *Nomads of the Himalayas.* Delhi: Sundeep Prakashan, Chapter 3.

Sherpa, M.N. (1993). Grass Roots in a Himalayan Kingdom. In E. Kemf (Ed.) *Protecting Indigenous Peoples in Protected Areas: The Law of the Mother.* San Francisco: Sierra Club Books, 45-51.

Singh, C. (1989). Rehabilitation and the Right to Property. In W.J. Fernandes & E.G. Thukral (Eds.), *Development, Displacement, and Rehabilitation.* Delhi: Indian Social Institute, 91-103.

Singh, V.B. (1989b). Elephant in NW. India (Uttar Pradesh). *Cheetal,* *28,* 39-43.

Singh, K.S. (1992). *People of India: An Introduction.* National Series Volume 1, Anthropological Survey of India. Calcutta: Laurens and Co.

Singh, S. (1994). *Biodiversity Conversation through Ecodevelopment: A Conceptual Framework.* Mussoorie: Centre for Microplanning and Regional Studies, Lal Bahadur Shastri National Academy of Administration.

Singh, R.P., Bhattacharya, P., Roy, S.B., Yadav, G, & Mahapatra, S. (1995). Community Forestry: A Case Study from Kudada, Bihar. In S.B. Roy (ed.), *Experiences from Participatory Forest Management,* Delhi: Inter-India Publications, 144-158.

SPRIA, & RLEK (1993). *Report on National Workshop on Declining Access to and Control over Natural Resources in National Parks and Sanctuaries.* Sponsored by Society for Participatory Research in Asia (SPRIA) and Rural Litigation Entitlement Kendra (RLEK), October 28-31, Dehra Dun.

SPWD (1993). *Joint Forest Management Update, 1993.* New Delhi: Society for Promotion of Wasteland Development.

Student Study Group (1995). *International Project on Community-Based Law Reform Proposals.* Delhi: Campus Law Centre, Faculty of Law, University of Delhi.

The Hindu (1996). Van Gujjars Rescue Baby Elephant. *The Hindu,* January 13.

Verma, V.K. (1983). *Gujjar Resettlement Plan for Rajaji National Park.* Dehra Dun: Rajaji National Park.

Vijayan, V.S. (1991). *Keoladeo National Park Ecology Study: Final Report.* Bombay Natural History Society, Bombay.

VIKALP (1994). *Rajaji National Park: Conservation, Conflict, and People's Struggle.* Saharanpur: VIKALP.

Wasteland News (1995). A Visit to JFM Areas in West Bengal, *Wasteland News, IX(2),* 79-81.

Williams, G.R.C. (1874; reprinted 1992). *A Memoir of Dehra Dun.* Dehra Dun: Natraj Publishers.

Wilson, E.O. (1988). The Current State of Biodiversity. In E.O. Wilson (Ed.), *Biodiversity*. Washington: National Academy Press.

Wood, D. (1995) Thou Shalt Conserve... *Down to Earth, 4(15),* 31-38.

World Bank (1993). *India Policies and Issues in the Forestry Sector.* Report no. 10965, Washington, DC: World Bank.

World Wildlife Fund-India (1989). *World Wildlife Fund for Nature-India, 1969-1989.* New Delhi: WWF-I.

Appendix I

The World of the Van Gujjars
- *Pernille Gooch*

A workshop with a difference

In February 1996 - less than four years after the Van Gujjars had been threatened with eviction from their forest, I was invited to partake in a workshop in Dehradun presenting the Van Gujjars Management Plan for Rajaji National Park. This was the manifestation of something that went beyond my wildest hopes. I have worked and lived with the Van Gujjars off and on since 1987. I have seen them marginalised, extremely worried over the future, and fighting what I feared was an uneven struggle for survival. I have during that time come to feel very strongly for all those Van Gujjars, men, women and children, I came to know and who received me, a stranger, with so much hospitality. I have further attained a deep respect for the Van Gujjars as I have seen them defying the forces that were trying to incorporate them as second class citizens in the mainstream society. They have kept their pride and a relative sense of independence, stayed in the forest in winters and walked up to the alpine meadows in summers with their herds of buffaloes.

Seemingly they maintained status quo but this does not necessarily imply stagnancy. For the Van Gujjars holding on to their particular way of life in spite of all adversities has been a major feat demonstrating their ability to adapt to changing circumstances. Still 'circumstances' might prove overwhelming and I feared that ultimately the Van Gujjars would be forced by a unilinear development process to become sedentary and loose their unique way of life. That would mean that one more of the traditional lifestyles adapting to the Himalayan ecology would disappear. As transhumant pastoralists the Van Gujjars have traditionally been part of larger socio-ecological systems made up by human communities interacting with the natural environment and with each other. Such systems of interdependence have evolved over long periods of time. But all knowledge on how to survive under those very specific circumstances is kept in

253

the living memory of people like the Van Gujjars and it would quickly disappear if they should no longer be allowed to continue with their way of life. In 1992 conservation policies and the proposed Rajaji National Park seemed to spell the end of the Van Gujjars pastorals in the area. This was the ' circumstances' to which it was not possible just to adapt and then get on with life. At that time I met Van Gujjars families in hiding in the forest outside the park, despairing because they did not know if they had a future. Now, we know they do.

The workshop to discuss the Van Gujjar Management Plan for the Rajaji area was held at the National Petroleum Institute and the inauguration ceremony took place in the large central hall with room for more than 400 people. Around 11 o'clock the hall started filling up with people. But what one saw here was not the usual conference attenders. The people who started pouring in were bearded men in *loongi* and *kurta* with rough blankets in soft red colours thrown over their shoulders and a smaller number of women also wrapped in homespun blankets. Several hundred Van Gujjars had come out of their forests and fully aware of their importance at this occasion they started filling the hall by occupying the front rows first. Quite obviously they had come with hope in their hearts and with a new confidence. It was the Van Gujjars who dominated the picture. There were also more conventional looking conference-goers present but they had to find their seats at the back of the hall.

.This was how the Van Gujjar leader, Mustooq Lambardar, presented his people:

"I am a Van Gujjar and I am here to tell you from where the Van Gujjars came and what we do for our living and what our ideology is. I congratulate all [of my people] because you have given us an opportunity to show everybody what the Van Gujjars are. There are many Gujjars in India but I speak only for the *Van* [forest] Gujjars, those Gujjars who have looked after the forest over the years, who have sustained a symbiotic relationship with the forest and who are the ones who have continued to stay in the forest. The Van Gujjars are my brothers and sisters. [Our] relationship with the forest is not a new one. It has been established over the centuries and it is characterised by looking after the trees and looking after the buffaloes. All the Van Gujjars share their feeling that if one harms the forest we would want to fight with that enemy. But, unfortunately, this power is not with us but only with the forest department. The forest department is making this forest into a national park which includes the trees and the animals *but nobody thought of including the Van Gujjars."*

No, strangely enough, there was the state contemplating making a national park to conserve wildlife and trees but earlier 'nobody thought of including the Van Gujjars'. This is now changing and the van Gujjars have presented their proposal for managing Rajaji National Park and I think they will make the most knowledgeable managers possible. Nobody knows the forest and the area as they do who have lived with it for generations.

A Changing View on Development.

I had visited the area for the first time in late 1987 to look into the situation of the Van Gujjars and plan for a larger field-work. What I found during my short stay in the Dehradun area at that time was a community living at the fringe of Indian society. They were in the forest and discourses about nature conservation and forest policies deciding their future as well as the future of other forest dwellers were being held in places from which their voices were barred. Their voices seldom reached further than local arenas, spatially as well as typically defined by representatives of the Forest Department.

In a study from 1986 Amir Hassan writes, 'the U.P. Gujjars... are an obscure group which was not only studied [by social scientists] but, strange as it may seem, has remained outside the planning process and thus has been largely deprived of the fruits of development during the last 32 years of our planning.' (Hassan 1986: xiii). Where other presumably backward and disadvantaged Indian communities have been listed as 'Scheduled Tribals' (S/T) or 'Scheduled Caste' (S/C) in order to be the beneficiaries of obtaining discrimination, the nomadic Gujjars of UP were not included in either list. As people on the move without an address they had at that time had no option being included in the voting list either and were consequently literally non-existent as Indian citizens.

During my next visit in 1990 I found that their life in the forest as well as during transhumance was becoming increasingly contradictory and problematic. They told of how they felt pressurised by forest regulations and by the arbitrary treatment they received from low forest official; and of the middlemen and dairy owners who exploited them when they sold their milk. Still, it seemed the best of life possible and viable alternatives appeared to be non-existent so the Gujjars tried to make the most of the situation and to maintain what they term '*Gujjari ki duwari*', the world of Gujjars by mainly keeping to themselves and quietly continuing with their own way of life in the forest.

When I visited the area in the winter of 1993-'94 I found to my surprise and utter delight that the situation of the Van Gujjars had become greatly transformed. Rural Litigation and Entitlement Kendra (RLEK) had undertaken the mission of empowering the community and had now started new development projects, in cooperation with the Van Gujjars themselves, meant to strengthen and intensify their specific pastoral lifestyle. These are the kind of development projects that the Van Gujjars had asked for during Nehru's rule, when he had vowed to help them, but which did not materialize at that time. Because of these interventions the Van Gujjars had a self-help scheme for selling their milk as well as programmes for veterinary and human health services. There was also an innovative literacy campaign with a set of primers taking their examples and vocabulary from the Van Gujjars' own life-style .

I visited the forest when the literacy campaign had just started and there are two episodes that I specially remember: One was a class of new learners at one of the Van Gujjar camps. It was held outside in a clearing with dense forest all around. The pupils, a mixture of men and women, girls and boys were learning to spell a new word "*Haque*" (right) and the teacher was asking "which right?" and the answer came loud and without hesitation, "*jungle ka haque*" (forest right).

The other was a father watching a lesson in another part of the forest. He was sitting down holding a young baby in his arms. The lesson was on one of the chapters in the primer which is about "hamaro jungle" (our forest), and as the lesson went on the father kept humming again and again to his child "hamaro jungle, hamaro jungle....".

Apart from that the Van Gujjars are now included in the list of S/T as well as in the voting lists. They have been accepted as Indian citizens at last.

Religion

Seeing the father in the forest with his young baby we understand how important the forest- being that which surrounds them and their daily activities- is for the Van Gujjars and how this feeling is acquired from early infancy. 'We are *jungli log*',- literally forest people or forest dwellers - is what they tell again and again when they try to either clarify what distinguishes them from surrounding populations or to stake their claims to the forest.

According to the Van Gujjars their life has been in the forests since ancient times. This is also told in their myths. Although nominally Muslims, the Van Gujjars must be seen as juxtaposed between two traditions, Hinduism and Islam. Originally the Van Gujjars were Hindus but at some time during the history of Mughal reign over India - most likely at the time of the emperor Auranzeb - they were converted to Islam. As part of their special identity the Van Gujjars have maintained a separate religious profile syncretizing Hinduism and Islam as well as showing influence from popular Sufism. Hasan (1986: 63) states that the [Van] Gujjars 'are an exceptional group [among Indian Muslims] in as much as they have ... made little effort to ashrafize [Islamizise] themselves.' Their Hindu roots are also clearly discernible in their vegetarian habits. Pastoral systems are specialized toward milk production and are often carnivorous as well because unproductive animals will be produced. Here, the Van Gujjars are unique in combining pastoralism with vegetarianism. **They firmly believe that , ' the buffalo provided us with milk all her life and so she is like a mother**

to us. How can we kill and eat our mother?' Neither do they hunt. As we will see below when discussing kinship, the Van Gujjars have also kept an exogamous 'gotra' system as reminiscence of their Hindu past.

The Van Gujjars speak of God by the name *Allah* or as *Kudrat* (Nature). *Kudrat* created the forest, people and everything else in it. When talking of 'ownership' of forest the Van Gujjars often make statements like, 'The forest belongs to God so how can the State or anybody else claim that it is their property?' The influence from Sufism may be seen in the tolerant attitude the Van Gujjars demonstrate towards other religions. When discussing religion, the Van Gujjars would say, 'it is all the same God, the Muslim God, the Christian God or the God of the Hindus, it is only the names that are different'. A famous Sufi story tells of a great Sufi who prostrated himself in front of one of the goddess of a Hindu temple. All but one of his disciples deserted him for worshipping another God but Allah. Only one pupil followed him understanding 'that none exist save God'. This story was not told to me by the Van Gujjars but the influence of stories like this among them is clearly discernible when Athru, an old woman, tells that the Goddess at the Hindu temple in the forest should be venerated because she -- as well as her temple - is also part of the 'sanctity of God'. The Sufi influence stressing the 'unity of all' is also discernible in expressions like, 'all life is the same' or 'our life is the same as the life of the animal in the forest.' This is also one of the reasons given for not hunting because 'how can you take a life which is ultimately the same as your own!'

One myth about the ancient connection between the Van Gujjars and the forest is drawn from their Hindu tradition:

Rama fought the evil king, Ravana. In this battle his army is believed to consist of monkeys. But these were not real monkeys, they were really forest dwellers, *'jungli log'*. After he had defeated and killed Ravana and was going back to Ayodhya these forest dwellers asked Rama what the message was for them now, and for the Van Gujjars the message was clear, 'you should go back to your jungle, keep buffaloes and hence pass your life.' And so they have done.

Creating a genealogy going all the way back to the mythic times of Rama and claiming continuity with the forest as well as with its animals - in the form of the monkeys - ever since creates a sense of belonging to the forest as good as any. Here we also have a presentation of the three main elements of Van Gujjars life; people (Gujjar log) - buffaloes (baas) - and forest (jangal). The three are intrinsically bound together, and together they make a unity. The Gujjar jangal in its optimum shape is a forest with a rich diversity of trees, scrubs, creepers and grasses providing a variety of fodder during the whole stretch of the cold season - while in summers the buffaloes graze in the meadows at the edges of sub - alpine forests. As forest dwellers the Van Gujjars spend most of their life in the forest (the time spent out of it is mainly the time used for the migration to and from the mountains.

The Van Gujjar often stress that they are born in the forest, die in the forest and eventually get buried in the forest. One woman told of how she was to give birth to her eleventh child. She was no longer young and the pregnancy had been difficult. She was staying in her *dera* (camp) in the forest high up in the mountains and she was being assisted by a *dai* (midwife) who was a close female relative. The birth turned out to be very complicated and her relatives, fearing they would loose her, decided to take her to the hospital. The nearest hospital was almost a whole day's journey away but they managed to carry her out of the forest and get her in a taxi. Travelling in the taxi she seemed weaker and they feared she would pass away. They then got very worried and decided to turn back. If she was to die it should not happen in the big town. She should expire in the forest and get buried there. They got her back to the forest and luckily both she and her baby survived but this story demonstrates how the forest is considered the only right place even for dying.

Growing up in the forest

For the Van Gujjars the environment is in this way 'internalised' as a knowledge gained right from childhood and based on a community's actual practice of survival in its natural environment. The youngster stays in the forest and takes part in the work of the family. They start with easy tasks together with other family members, as carrying very light bundles of leaves, and for this they get praised. And so they gradually learn the art of survival of both animals and people in their forest environment. Something that struck me while staying with the Van Gujjars was the fact that girls and boys were equally treated in the families. I never saw that a small girl was less loved, had worse clothes, or was less well fed than her brothers. On the contrary it was often the youngest daughter who was the child most spoiled and most indulged in by her family and perhaps specially by her father. In the same way both boys and girls help from a very young age with the work in the forest. Youngsters of both sexes will be seen to accompany each other into the forest to fetch leaves or grass for the animals. While boys or men might be seen more often climbing trees to cut leaves for the animals it is not unusual to see a girl high up in a tree or even a young woman perched high up with her young child waiting below.

Walking with a small Van Gujjar child in the forests of the Shiwaliks, it will relate the trees and plants passed by as elements in a world that for the child is already full of meaning and shared by humans, animals and trees. This sharing also includes wildlife, something which is clearly discernible in the obvious joy of the child when he spots a wild deer in the form of a cheetal or a kakar. Trees are related according to their relevance as fodder for the animals. Here the child will relate different kind of leaves according to its own horizon and will tell how some leaves are like roti (bread) for the buffaloes, something with which to fill their belly, while others are like sweets, eaten mainly because of their taste. In this way an environment is known through living in it. Such a world we may term a 'life world' e.g. world about which we get knowledge directly and from within (as contrasted to theoretical knowledge about the world gained through studying. A knowledge that we acquire through looking at something - for example a forest - from the outside). This is the 'Gujjari ki duwari', the world of the Gujjars, that I mentioned above. The British forester and nature lover F.W.Champion, who worked in this part of the Shiwaliks in the twenties, also refers to a state of symbiosis between Van Gujjars and the wildlife. He tells of hunters coming from outside to shoot tigers in the jungle using a young buffalo as bait. But, says Champion, the tigers who are familiar with the ways of the Van Gujjars, 'the jungle herdsmen', should know that the Gujjars would never leave a calf unguarded in this way and a 'wise' tiger should consequently understand that such an animal - contrary to buffaloes kept by the Van Gujjars - does not belong to the forest and avoid it. (Champion 1924;126)

This intimate knowledge gained through the ages of how to survive in the forest and in the mountains is expressed like this by Mustooq Lambardar :

"The origin of the Van Gujjars goes back thousands of years. We grew up in the jungle with our buffaloes. In the jungle we teach our children how to make food and there we teach them everything else of how to survive. Our birth is here in the jungle and the 'children' of the wild animals are born together with our children. For ages the Van Gujjars have gone to the forests in the Shiwaliks in the winters and during the hot days we go to the mountains and live there and we bring our children with us. We stay at heights of 16000 - 18000 feet. You may ask why we like to go to the mountains ? The answer is that our birth is in the jungle and we also love the mountains."

As we see it is the whole family among the Van Gujjar who partake in the nomadic life. Watching a group of Van Gujjars on the way up (or down) the hills we may detect them along one of the trails as a large and straggling band of people and animals winding and bending round the mountain. And in this colourful array of humans and animals there will be people of all ages from the oldest grandmother bent over her stick to the tiniest baby carried on the back of an older sibling or by one of the parents (among the Van Gujjars, fathers may also carry their babies). For many other pastoral nomads - as the Gaddis, sheep herders from Himachal Pradesh - livestock rearing is combined with agriculture and only part of the family participate in migration with the herds while the rest of the family stay back and take care of their fields. A similar pattern will be found in the Seiss Alps or in the North of Scandinavia. But the Van Gujjars have no other occupation than pastoralism and so the whole family is involved in it.

Transhumance

Being part of a regional system

What has dominated discussion round the Van Gujjars and their use of the natural environment has been their close relations with the forests of the foothills. But as pastoral nomads they utilise a much larger region. The whole logic behind the community management plan for the Van Gujjars in the Rajaji area builds on the fact that the Van Gujjars leave the area during the drought of the summer giving it a chance of recuperating during the monsoon before they return in Autumn.

The region traversed by Van Gujjars during seasonal migrations is huge. It consists of the hilly region of Uttar Pradesh and the districts of Himachal Pradesh[1]. While there are many other groups of nomadic or semi-nomadic Gujjars in the western part of the Himalayas the nomadic Van Gujjars in the region I have outlined here constitute a separate population, recognising a common identity and linked together by ties of kinship.

This region then with its rich ecological variations is the physical arena for the migratory life of the Van Gujjars. Whereas in agro-cultivation humans create (and control) their own miniature ecosystems, such control is not possible for people involved in herding of livestock. Pastoralists have to conform to the natural environment. One way of doing that is by transhumance, defined as seasonal oscillations of herds

---

[1] Gujjars are also found in the western parts of Himachal Pradesh as well as a large population in Jammu and Kashmir and many Gujjars from those parts spend winters in the Punjabi plains

between low land areas and mountain pastures. It is the physical environment that shapes the seasonal migration pattern. As transhumant pastoralists the Van Gujjars utilise ecosystems at different altitudes according to the availability of fodder and water. In summers the Van Gujjars migrate to the pine and spruce forests in the mountains and the buffaloes are let to graze in the alpine meadows, the bugyals.

In this way the Van Gujjars have adapted their way of life to the changes in natural ecosystems being at each time of the year in the horizontal ecological zone optimising survival for them and their herds. Such a way of life is highly specialised and demands an intricate knowledge of the environment. Above we discussed how such an environment known through practice constitutes a 'life world' for the people using it. Looking at it like that we see that for a community of agriculturists the village and its surrounding will make up the life world (as well as the local ecosystem) because this is the environment being used - as well as being 'created and controlled'. For a pastoral nomadic people it is the whole region traversed during the yearly migratory cycle that will be of environmental relevance as a 'life world'. It is within this setting that practice e.g. pastoralism, is enacted. As we will see this way of looking at the environment fits well with the way the Van Gujjars themselves look at it. They see the whole region as one large system where everything is interconnected.

They tell that in summers the force of 'life' itself moves from the dried out forest in the foothills up into the new fresh pastures in the mountains. After summer 'life' moves the other way back into the lowland where everything will be fresh and new after the monsoon while the 'life' in the alpine meadows will die covered in snow. In this system the Van Gujjars and their buffaloes follow the cycle of life on its annual movements. This is the miracle of Kudrat. As we saw the name Kudrat is used simultaneously for Nature and for God. And so it is the miracle of the God who is Nature that there is always life in the form of green fodder somewhere in the larger system. In this way the Van Gujjars survive following natural changes.

This is beautifully described by Noor Jamal from Timli range:

"When we come down after the summer we find everything looking (as if it was all) one, all is green and you will not find a single tree that is cut. When we leave (the foothills in the late spring) then the forest is dry and pale. The hot wind (loo) will fire from this side- from that side , blow - blow. There will be life in the hills. That will live and this will die. That will be Kudrat's miracle. Then afterwards that will die and this will live (indicating the transfer of life from the mountains up there to the forest down here). When we go down the leaves will keep drying in the hills. There it will snow and leaves and grass will become buried."

(What happens to the wild animals when the forest dries up?)

"It is like this that light summer rains will give life to jungle grass (he shows strands of green grass growing at the edge of the rau) and on that the animals will live. *Kudrat* provides only grass enough for the wild animals to live on (in summers), not for Gujjars. For Gujjars it will be in the hills."

Another man expressed it like this, "It is good if we can go to the mountains because then the trees have peace and when we come back they will all be green. If we stayed down here for all twelve months there would be no green grass and no leaves left. We know that. We love the forest and that is why we leave it so it can be restored." Here we see the Van Gujjars stressing the ecological adaptability of their own nomadic way of life.

Earlier the Forest Department encouraged the Van Gujjars to become sedentary so they would stay in the foothills all year round. This was supposed to be the first step towards settling them completely. As it turned out it was a 'development' project that backfired, because it only intensified the situation of degradation in the forest. That the Van Gujjars themselves, provided the choice, understand the importance of transhumance and prefer it is clearly demonstrated by the fact that nearly all Van Gujjar families went to the mountains this summer (1996). Included in this massive wave of summer migration was a number of families who had otherwise given up nomadism years ago. The reason given for this new development was that they now fell relatively safe in their forest in the Shiwaliks and know that they will be allowed to return in the autumn.

Above all the Van Gujjar point of view on transhumance was expressed in quite sophisticated ways. Most Van Gujjars, when asked why they go to the mountains in summers, came out with more simple answers. A very old woman expressed it like this, "we go to the mountains because we love the coolness there in the summers. Neither we nor our animals are used to the heat of the plains in the hot season. If we were forced to stay back we would be sure to die." This old woman had never experienced a summer in the low lands. She had gone to the high mountains in Uttarkashi every year of her life and spend the hot seasons at an altitude of 3000 metres. now in her seventies she still walked proudly all the way with her loaded horses accompanied by her daughter in laws, sons and numerous grandchildren.

Social organization during migration

Pastoralism is, as we saw, a family venture and as nomadic people, the Van Gujjars do not have a house or a piece of land to symbolise belonging. Belonging is maintained by being part of a dera. Like for most pastoralists the household is the basic unit of Van Gujjar society, functioning both as units of production and consumption. The physical appearance of these units is manifested in the dera, often translated as 'camp' but a dera is both the 'home' of an individual household, the people sharing the same cooking fire, and the migrating household. In this way the dera symbolizes belonging and home and when stationary the dera of this or that person is where you go to visit. But because it is the home of a nomad it migrates with him or her and as such it includes everything that moves.

Deras are usually represented by their male head. Still this is a truism with modifications and in several of the families that I know personally the power within the family lies with the eldest woman [2] and the dera may even be represented by her name as the dera of Noorbibi, Tajbibi, etc. Traditionally women have a strong position among the Van Gujjars and while the responsibility for production is shared by all family members it is the female head who takes charge over the economy of the family. All money is given to her and she distributes it and negotiates its use with the rest of the family. I have seen families sitting down together before migrating in early summer to discuss the economy on the way. The woman sits in the middle surrounded by her family. She has the money, takes it out, and for each expected cost she leads the discussion on how best to manage it and then puts aside a sum of money.

During transhumance a small group of deras migrate together. The structure of such groups might vary from year to year as they are kept together by ties of personal friendship as well as by kinship. The groups have to be relatively small not to restrict mobility and still large enough for mutual security against robbers and other dangers.

Life during migrations is hurried (or as hurried as is possible considering the slow pace of buffaloes). At the *Padavs*, (encampment sites), the families take shelter under their *Tambus*, (tents) made of just a simple sheet of canvas or black plastic, if it rains. When the weather is dry they sleep under the open sky. But it is the buffaloes that determine the pattern of the migration and not the people, so already at 2 O'clock in the morning the buffaloes might get restless and hungry and they will then start moving. It might be possible to restrain them for a while but ultimately the herd will be on its way and the people will have no option but to follow.

Part of the family with the youngest children and the beasts of burden (horses or bulls) will stay back until morning when they will pull down the camp and load the pack animals

---

[2] This is, of course, the case where there are no male heads as in a case of an elderly widow whose sons have not yet attained sufficient authority to become heads of their families, but it may also happen in families where the man has lost power because of old age or because his wife has attained more 'social power' than he has.

before joining the trail.. usually men and youngsters of both sexes walk with the buffaloes while women, again accompanied by a mixture of youths, take the young children and the pack animals. But this pattern is not cemented and in actuality family members of both sexes might join either group. This shows the flexibility in the social organisation of the Van Gujjars. What people actually do depends on the labour power available for the family at a certain time as well as on personal preferences. This may be demonstrated by the case of Sen bibi and Lal Sen. Sen bibi, a middle-aged woman says she prefers the slow pace of the buffaloes and she walks with her small herd of seven animals while the other herds in her migration group are headed by men. Luckily, her husband has very strong personal ties to his bulls so he takes the luggage and accompanies the women of their migrating group.

The halts are mainly the same from year to year but each day both men and women of the group meet to discuss the route for the following day. In this way decisions are taken together by both sexes and when asked about leaders during the migration people would say 'we do not have a leader. We decide together'. One old woman went as far as to say, 'when we go to Dehra Dun to the market we do not have a leader. Then why should we need a leader when going to the hills?' It is, of course, necessary that somebody walks first and its usually the man with the largest herd but he has to negotiate routes and halts with the rest of the group. As the women point out it is in actuality the old buffaloes and cows who lead the migration because they know the way and in spring they are eager to leave the dried out forests in the foot hills and eat the new grass in the mountains. In the autumn again they want to return to the forest with its new leaves.

The buffaloes walk slowly and they can only move in early morning or in the evenings when it is cool so by midmorning the new camp site must be reached. The rest of the group walk much faster with the pack animals and although starting hours later they will join shortly after and start putting up the camp and preparing tea and lunch of roti and milk. During migration the Van Gujjars may exchange milk or milk products for vegetables with the settled populations which gives added nutrition to the meals. Otherwise milk yields are small during this period and most is given to the calves or the children. A considerable part is also used to grease the palms of minor forests officials on the way. At some halts there will be little fodder for the animals and it will not be possible to tarry and make food or tea and at such times milk, drunk directly from the udder, will be the only nourishment available. Stretches covered during a day's march are short, usually between 8-15 kms. Migrations usually take around a month but as the Van Gujjars spread out over a very large area during summers there are of course variations in the time spent on the trail for different groups.

Relationship with settled populations

Some of the groups migrate through Chakrata and Tuini and they cross the Tons river and spend summers in the eastern part of Himachal Pradesh. In the Shimla hills there is a form of symbiotic relationship between the pastoralists and the villagers. Van Gujjars graze their animals on clearings in the spruce and pine forests and the settled populations live on the slopes below. In Himachal apple orchards have proved to be a very lucrative business and many of the villagers are now very well off. Most of them keep a couple of milk cows just for family needs but apart from that they do not engage themselves in livestock rearing and milk production. This leaves a 'niche' open for the Gujjars to produce milk for a local market in summers. That the Van Gujjars are very important for milk production here is evident from the fact that the owners of the small tea stalls apologize for having to serve 'red'(milkless) tea after the Van Gujjars have left in autumn. Part of the milk is also made into khoya (a mass of boiled down milk from which Indian traditional sweets are made) directly in the forest and sold to the sweet dealers in Shimla.

Other groups stay in Uttar Pradesh and migrate to Uttarkashi or into the Garhwal Himalayas (a smaller portion also migrate to the Kumaon Hills). Here they camp in the sub-alpine forests and graze their animals in the Bugyals above. These areas have some of the most beautiful scenarios of the world and there are places which attract tourists as well as pilgrims. All those people need milk and milk products and that is provided by the Van Gujjars. This kind of interdependence is part of old patterns where human communities have interacted both with the natural environment and with each other. In this system the Van Gujjars have specialized as the producers of dairy products using the 'natural' environment for such a production while the settled farmers have used their small terraced fields to produce food direct for human consumption. Returning to the concept of regional system we find that the Van Gujjars, utilizing natural resources on non-cultivable land, are an intrinsic part of a regional exchange system. The Van Gujjars are, through their specialized production, in direct interaction with the natural components trees or grass of the environment. Through the practice of milk pastoralism, forest roughage, indigestible by humans, is consumed by ruminants and delivered in a highly digestible form. According to Ingold the advantage of milk pastoralism lies in the fact that it provides for greater carrying capacity of uncultivated pasture land than does carnivorous pastoralism (quoted in Ellen 1991: 75). By practicing a form of 'human parasitism' the pastoralists out compete the natural offspring of the milk animal and they are able to partake in the food chain at an earlier stage than are pastoralists who kill their herds for meat (ibid.). So what it comes down to is that Van Gujjars through the 'medium' of buffaloes change forest biomass into milk products. The 'natural' and 'unpolluted' component in the production is a fact that is now being used with advantage as Gujjars have started supplying milk to an environmental and health conscious urban population in the large towns within the region like Dehra Dun and Hardwar. It should also be noted that while the term 'pastoralism' is often used in a 'slovenly fashion' to include many people only partially dependent on livestock for survival (Ellen 1991: 271) this is not the case for the Van Gujjars who are entirely dependent on their herds for survival.

A distinction is usually made between a market oriented pastoralism and one based on subsistence. From the discussion above we see that the Van Gujjars produce mainly for a market but this is a truth with modifications as milk products are simultaneously used for subsistence and a considerable amount of milk is kept for domestic use. This actually makes up an important part of the food intake as most of the protein is consumed in the form of milk or milk products. Some of the more wealthy families with which I stayed told that they used 10-12 litres of milk a day for the household and for the guests. The Van Gujjars are very hospitable and it is considered an honour to entertain guests. So for families who want to gain respect, *izzat*, from their community it is important to have enough milk to serve to visitors. Even poor households will use a substantial part of their milk production for guests rather than selling it and earning money from it. Milk and milk products are also used as presentations at weddings and other ceremonies where all guests will arrive carrying a jug of milk or a slab of butter and quietly leave it at the back door.

The herds

For reasons of ecological adaptability it is important to look at the particular breed of buffaloes kept by the Van Gujjars. The Gujjars' buffaloes are an "indigenous" breed of small hardy animals kept by them for generations and well adapted to their special lifestyle. They are efficient in changing forest roughage into milk and they are better at walking for long hours in a rough terrain than are other breeds. They also have a much higher fat content than other buffaloes, around 10%. A good milker will produce about twelve liters of milk a day on forest roughage and a small supplement of bran and oilcakes. The average yield in a herd lies around two to four liters a day but then it should be taken into consideration that the calf will stay with her mother during day-time and drink from her so in reality more milk is produced.

Yields are highest in winter as most calves are born in autumn but the production drops quickly in February - March when the forest starts drying up. Summers yields are lower than winter yields but they have a higher fat percentage.

The Van Gujjars had developed their own breeding programme for improving their herds and for keeping them healthy. They usually keep the bulls from the best milkers and they try to exchange bulls or male calves with deras (camps) quite far away. They prefer changing bulls frequently and they never buy bulls from outside as they think such bulls will 'weaken' their own breed of buffaloes. This means that we here have a unique 'bank' of animal genes 'walking around' in the shape of the Van Gujjar herds. It also entails that any development program being lured by the possibility of higher yields by cross breeding could fast prove counterproductive and what would be produced would be animals completely unsuitable for the life of the Van Gujjars. The Van Gujjars themselves realize this which again has been very eloquently expressed by Mustooq Lambardar : [?]

"Those buffaloes that give 50 kgs of milk if they came into the jungle they would not even be able to walk one km. But our buffaloes who can walk so far, who stay hungry when there is no food they have [the right] nourishment. If you bring a buffalo from the town you need two people to run behind it. When you walk up into the mountains when it is snowing such buffaloes will never do it because they do not have the proper strength. You can't even take them halfway because they will be all dead by that time. Such a buffalo will just stand there looking at you. It is like when you bring a person from Bombay and you take him up into the mountain. He can't walk by himself and you have to take him up in a carrying chair. Even if our buffalo does not get food for two three days on the migration she still gives two three litres of milk but the ones from the town would not give any milk if they had to stay hungry. There is too much work involved in getting the milk from them. It takes two people to tie it, to wash it, to make it run. Our buffaloes tolerate a lot, thirst and cold and hunger and climbing."

I think the case against cross breeding could not be put any stronger and more persuasively than this.

Like for most pastoral people there are strong ties between people and their herds and buffaloes are considered as friends with histories and personalities. One Gujjar story tells that, "when God created the buffalo he gave it a very large belly and the buffaloe wondered and asked God, "How am I to fill up this large belly?" But God answered, "Do not worry I will attach something that will make man take care of your needs." Because there is a system of reciprocity between animals and man it is of importance how animals are treated by their human guardians. A woman tells that 'male calves are sold at one year old to be used as draught animals down in the plains', 'but', she continues, ' this is not unproblematic because the life of the buffalo bull will be miserable, he will be forced to pull heavy loads and he will curse the person who put him into such misery. This again will have repercussions for the people and the curse of the buffalo bull will affect them as well as all their female buffaloes'. The Van Gujjars live in a world of reciprocity e.g. a world where every action you take will have consequences that ultimately affect you in some way or other. In such a world all other beings must by necessity be treated with kindness and care.

As for all pastoralists the possibilities inherent in the natural environment are perceived through 'the-eyes-of-the-buffalo'. When passing through a particular landscape on the downward migration it was stated by an old man, 'this is Gujjar land, good grass, running water, the shadow of trees and no agricultural land close by for the buffaloes to wander into' indicating that a landscape that is good for the buffaloes is good for the Van Gujjars. Forests are evaluated after the quality and proportion of its fodder trees. A good forest is one that has a large proportion of 'trees good for buffaloes to eat' species as: *bakli, sain,* (here some claim that *sain* is the best and other prefer *bakli), papri, Sandan, panchara, and maljan.*

Another issue to take into account when discussing herds is that most Van Gujjars have their complete wealth - well as their only 'mean of production' -invested in them. Such wealth may be characterized as 'riches on hooves' (Ingold 1988). As we can understand such 'riches' are a very 'fragile kind of property, and an epidemic or natural calamities may decimate even a large herd under a short spell of time. This means that the Van Gujjars can never be sure about the future. The animals may die or they may stop breeding for some reason or other. In both cases the production of milk will dwindle and the future of the family be threatened. In such cases it is the traditional, internal solidarity of the community that will enable a family, struck by disaster, to survive as pastoralists. In this way the Van Gujjars have always had to co-operate in order to counter the risks involved in living under harsh ecological conditions.

Kinship

Like other tribal societies the Van Gujjars construct their social structure round the kinship system. As we will see it is to a great extent kinship that defines how people relate to each other, and how they acquire possession of the basic means, necessary for survival as migratory pastoralists: animals, rights to pasture, and household utensils.

Above I discussed how the Van Gujjars have kept the exogamous *gotra* (clan) system as a reminiscence of their Hindu ancestry. This means that all Van Gujjars are divided into a number of named *gotras*. Some of those are *Kasana, Lodha, Baniya, Chechi, Khatana, Pashwal, Dinda*. The *gotra* system is patrilinear which means a person born into the gotra of his or her father. The members of the *gotra* trace relations through a fictitious, common ancestor. In some cases we have a narrative of the origin of the *gotra* mentioning a mythic ancestor while in others there is no such elaborated myth.

I have here chosen to narrate the myth of 'how the Kasana *gotra* came into existence'. This story also gives a valuable insight into some of the points concerning family life that I will discuss below.

Origin of the Kasana gotra:

Asu Chechi was the ancestor of the Chechi gotra. From his family the Gujjars trace also the origin of the Kasanas. The story goes like this:

Asu Chechi was a very rich man, having many buffaloes and lots of wealth. And he had two beautiful young daughters and four [other daughters] who were very small when this story took place. His only concern was that there was nobody to manage his wealth and buffaloes, and he himself was too old to manage them all.

Near his *dera* in the forest, there were many [forest] contractors working who had engaged labourers from a foreign land for cutting wood. Two labourers named Jeengad and Ghasseela, who were Hindus, used to visit Asu Chechi's dera from time to time to have lassi [buttermilk] and whey. One fine day Chechi got an idea. He told Ghasseela and Jeengad that they could have as much lassie and whey as they could drink if they would do him a favour in return. He then asked them to fetch firewood for cooking and fodder for the buffaloes from the forest.

Now Ghasseela and Jeengad were very happy as they could ask for as much buttermilk and whey as they wanted. So they set out to get firewood and fodder.

After spending the whole afternoon in the forest collecting firewood and fodder they returned in the evening to Asu Chechi's dera. When Asu Chechi saw the firewood and the fodder he was astonished to see how much firewood and fodder Jeengad and Ghasseela had collected and carried on their backs at one time. When the fodder was given to the buffaloes it was more than they could eat and as a result much was left uneaten by them. Asu Chechi could not help feeling that these two would prove very useful in helping him take care of his buffaloes and managing his work.

Asu Chechi then got what he thought was an excellent idea. He discussed it with his wife and she agreed with him. The next day when Ghasseela and Jeengad came to Chechi's dera to give him firewood and fodder, and to take buttermilk and whey, Asu Chechi asked them a question. He asked them if they would mind being his son-in-laws and stay with him in his home. He explained to them that he had two young daughters. Jeengad and Ghaseela got very excited and they said 'yes' and then they added that they knew they had to have a *gotra* [in order to become Gujjars and marry the daughters] and so they became the first of the Kasanas.

In this narrative we see how the mythic Asu Chechi manages to get husbands for his daughters as well as securing labour for his *dera*. For the marriage to take place Jeengad and Ghaseela had to become Gujjars and they needed a *gotra* affiliation. A marriage cannot take place unless the participants represent a *gotra*. The Van Gujjars are exogamous and inter-marriages among members of the same *gotra* is strictly avoided.

In marriage regulation ones' own *gotra* is the only one to be avoided. That means a Gujjar boy may marry into the *gotra* of his mother or of one of his grandmothers. This is not only permissible but also considered desirable and cousin marriages are preferred but because of gotra-exogamy they have to be of the cross-cousin type and not the parallel type commonly in use among Muslims. In a community like the Van Gujjars, where survival for an individual family hit by calamities, may be completely depend on assistance from kin, it is of course important to have as many close knit relationships as possible on which one may depend in the case of need. The strengthening of already existing networks builds up strong bilateral ties of solidarity and is consequently a very good and sound strategy. The Van Gujjars also stress that it is often the old people of the family who insists on marriages between close family members.

This happened in the case of Roshanbibi. She was getting old and thought she might not be around much longer but while she was 'still in this world' she wanted to make sure that the members of her family would 'stick together' and keep helping each other also after she was gone. She then set out to arrange marriages between her grandchildren. As they had to be of the exogamous cross-cousin type it would have to be the children of her daughter, Noorbibi, marrying the children of her son, Kasim. Noorbibi also wanted to maintain strong ties with her brother's *dera*. Here, we see women's strategies behind the marriage negotiations, and as mentioned, the active part in marriage arrangements of this type. The preferred way of contracting marriages is by exchange, *satta*, e.g. a daughter for a daughter. In this case that means the *dera*s of Kasim and Noorbibi exchanging daughters and two marriages being secured.

Patrilocality, e.g. a married woman moves to her husband's *dera*, is the rule (but as we saw in the myth above is not a rule without exceptions). In my example that resulted in Kasim's daughter going to stay with her, *phuphi* (father's sister) and Noorbibi's daughter to her *mammi* (mother's brother's wife). This means that both girls go to stay in *dera*s with people they have known since early childhood, and one reason given by Van Gujjars for preferring this kind of marriage association is that they hope their daughter will be well treated among her close kin after she has married and moved away. Still, whenever a married woman moves away or not she maintains strong links with her own kin all through life and affinitive links based on the sister/brother relationship from very powerful networks for mutual assistance.

In the narrative of the mythic Asu Chechi we saw a different pattern as it was Jeengad and Ghasseela who came to stay in the *dera* of their father-in-law. In this way Asu Chechi's family has many buffaloes, daughters of the right age to marry but no sons (or only very young sons) to help with the work. In this case it is not necessary to make an exchange for the property from the father-in-law. We also see that the cross-cousin model is not always possible and then eligible brides will be sought through the kinship networks.

If exchange is impossible for example in families having more sons than daughters the last resort is to give bride price [3] and although it is much used it is still an issue of controversy and it is considered dishonourable. Feroz, the father of seven daughters and only three sons, tells proudly that he has not taken money for any of his daughters and he gives as a reason that 'four things are forbidden in our religion: 'gambling, drinking, theft, and selling of girls'. In a provocative question I

---

[3] I am using the word "bride price" here as in this case it is a question of a monetary transaction from the groom's family to the father of the bride and the expression "selling of girls" is often used by Gujjars criticising the phenomenon.

asked if it was not very good for a family .to get many daughters and the losers in this structure are the families having many sons and few or no daughters. For them it is very difficult to secure marriages for all sons.

The Gujjars also practice levirate as a widow is remarried to one of the brothers of her deceased husband. The Van Gujjars may also divorce and remarry and in most cases the marriages are broken because a girl does not like to stay at her in-laws place. She will then move back home and negotiation will be started to free the girl from the marriage and get her married somewhere else that is more to her liking[4].

Cattle-wealth

An important aspect of herds-management is the ownership of animals. It is often said about herds of pastoralist that they are owned either by the household or by the male head. Something which in reality comes to more or less the same and means that it is men - mainly older men - who has the power over the cattle-wealth of a family. In the case of Van Gujjars we will find a much more complex picture when looking into the ownership pattern of actual herds. Here we find that while the herds are maintained collectively by a household the animals are individual property and may be owned by either men, women or boys.

There are number of ways through which individuals attain the ownership of animals. A boy gets his first buffaloes when he is circumcised as a very young child, ideally one from his father and one from his mother. Weddings are also occasions for the presentation of cattle. At his wedding a boy should get a buffalo from his parents and the bride gets a buffalo in *Mehr* from her in-laws. The latter should be a very good animal, preferably one of the best in the herd, and it should be giving milk. After the wedding the husband has to ask his wife's permission to take milk from that buffalo. This

---

[4] This is something that might prove very expensive as the husband's family will not let her go without a compensation that should be at least as much as they would have had to pay in bride-price.

custom is so important that people will say 'without' a buffalo in *Mehr* (its name should be told as part of the wedding ritual) there will be no wedding'. When a person establishes a separate *dera* he will receive a share of the herd of his parents. How many animals he will get depends on the size of the family herd at the time of separation which means that sons might not get equal shares. When the father dies his remaining herd will be shared by his sons.

Apart from getting a buffalo as *Mehr* a woman should also get animals from her own family. Ideally, a brother should give a buffalo with a young calf to each of his sisters after their marriage but if he is poor he will not be able to do this and may give only calves or goats or perhaps utensils or nothing at all. Women will tell that it all depends on his means and the emotional bond between him and his sister. Parents might also provide daughters with buffaloes. In several cases I was told that more buffaloes had been presented to sisters or daughters married to poor husband's with few buffaloes than to wealthier ones. Here, it was a question of securing a living for a daughter/sister and her family under circumstances that would be very difficult otherwise. There are many families who have received the main part of their livestock wealth from the wife's family. The fact that women own buffaloes helps to strengthen their position in the household as well as in the community. Backed by the ownership of a personal herd a wealthy woman might use it to negotiate a powerful position as she can present buffaloes back to members of her own lineage to secure their support.

I have mentioned how kinship networks function as important means for internal solidarity. This means that in cases where individuals or households are hit by disaster, threatening their survival, assistance will be given to ensure that they can continue as members of the group. To secure that, they have to have the basic means for survival within the complex of nomadic pastoralism-buffaloes, an access to pastures both in summer and winter, and the necessary household items.

The household of Yusuf Ali consists of himself, his wife Tajbibi and their five young children. Their small herd of four milk buffalo, two heifers and two calves were hit by disease and five animals died within a couple of days leaving one pregnant buffalo and the two calves. Survival with so few animals was not possible and the family had no milk either to sell or for the household. One cousin came with a cow and calf so there would be milk for the children. Tajbibi's brother presented a young buffalo and other relatives donated buffaloes and heifers. The herd of family was reconstructed and they could take part in the migration with the rest of their migration group. With the reconstruction of the herd came a reassertion of Van Gujjar identity as the family was kept within the context of pastoral management and not allowed to drop out.

Appendix II
Harmony Between Human Beings and Other Forms of Life
- Bharat Dogra

The relationship of human beings to other forms of life on earth has never received adequate attention, and animal rights activists have a very important role in reversing this neglect of animal rights.

There are two very important justifications for this. Firstly, forms of life other than human beings - mammals or birds, aquatic life or amphibian - experience pain as much as we do and it is our duty as the most powerful species on this planet to minimise this as much as possible. Secondly, we have our own self-interest involved. As various forms of life perform important roles in the maintenance of ecological balance we also serve our own interests if we are careful to protect various forms of life. This is not to deny that some efforts will always have to be made to control clearly harmful insects or other pests, but at a more general level the ecologically useful role of most forms of life has to be accepted.

These clear and simple justifications for respecting the rights of other forms of life can be supported by thousands of convincing examples of how useful the various species are and yet how much needless cruelty and injury we inflict on them. Yet, somehow, this campaign has failed to be very effective. At least for the sake of voiceless animals, birds and fish it is important to discover the reasons for this failure and try to take corrective steps.

To some extent, of course, this failure is part of a wider trend of growing insensitivity to pain and suffering around us. However, despite this constraint the justifications for giving other forms of life a better deal are so strong that it is possible to motivate even an insensitive society to take more significant steps to reduce the pain inflicted on them.

Unfortunately, despite the good intentions and dedication of many animal rights activists, some self-imposed distortions and restrictions have greatly hampered the ability of the movement to help the forms of life and involve a wider section of the society in this effort. One such distortion is the heavy bias in favor of a few majestic species such as the tiger. While recognizing the important ecological role of the tiger, from the point of view of maintaining ecological balance and reducing pain many times more could have been achieved if at least one tenth of the attention given to tigers was spared for fragile, vulnerable and humble animals such as frogs. Frogs are extremely useful in controlling harmful insects, and yet they have faced an alarming reduction in numbers in recent years, a mistake made worse by alarming, highly unjustified cruelties inflicted on this gentle and delightful species. India's indigenous bee species provides another example of a non - glamorous but extremely useful (particularly for pollination) form of life being entirely ignored even when high mortality and near extinction from some parts is reported, that too partly due to man-made causes such as the ill-considered introduction of exotic species.

This bias in favour of the more majestic and colourful animals at the cost of very useful but less attractive and glamorous ones distorts the basic considerations (of reducing pain and ecological disruption) of the movement and due to this distortion it becomes difficult to secure the involvement of a larger number of people.

A related distortion is the bias towards the relatively small protected area of parks and sanctuaries to the relative neglect of the importance of animal rights in other areas including ordinary forests, rural areas and to some extent even urban areas. If all types of animal life is considered together then many more of them live outside protected areas than within them. Their survival is influenced by many factors such as forest and agricultural policies, the extent to which pesticides and other poisonous chemicals are used in agricultural fields, what kind of dams are built on rivers, to what extent rivers get polluted, whether restrictions (voluntary or legal) exist against the indiscriminate killing of animals, etc. Breeding and disease control policies followed in the case of domesticated animals are also very important because wrong policies can easily lead to the replacement of indigenous, hardy, disease resistant breeds with new breeds not really suited to local conditions and more vulnerable to several diseases.

Thus, what is really needed is a concern for other forms of life everywhere on earth, and giving their rights and their problems a place in all relevant aspects of our development and planning, irrespective of whether this relates to agriculture, animal husbandry, forestry, water pollution, dams agrochemicals, etc. The overemphasis on a small area of sanctuaries and parks comes in the way of this wider perception of the problem and this leads to a false sense of our responsibility being over with the creation of protected areas.

Another serious distortion which has created a lot of avoidable controversy is that some animal rights activists unfortunately give priority to those ways of working which bring them in direct conflict with human beings, that too generally needy and poor people who themselves need some help in their survival struggle. This is most clearly seen in the efforts to evict villagers from places where some protected areas are being established. Surely there must be some ways for protecting wildlife other than those which call for displacement of people, particularly keeping in view the fact that agricultural fields also attract many types of birds. Yet somehow this form of protecting wild life has been preferred to the exclusion of others which can prosper on a protective relationship between villagers and other forms of life.

Another example of avoidable confrontation is that when some wrong practices on the part of pastoral people, fisherfolks or other such groups are noticed, some enthusiastic animal rights activists arrogantly blame such groups without pausing to think if these poor people are themselves the victims of wrong polices. Some of these communities have been victimized and alienated by development policies and programmes of recent decades, and it was their crisis of survival that pushed them to some destructive and cruel practices. The problem should be seen in this wider perspective so that the grievances of these people can also be taken up while persuading them to give up any practices that cause great pain and suffering to other forms of life.

After all, the aim is not to create controversies that do not take us anywhere but to actually achieve a reduction in the suffering caused to animals while also protecting the ecological balance. We cannot achieve this by antagoninsing those very groups of people who live nearest to wild animals, or whose own livelihood depends on other forms of life. Animal rights activists have to work with these people, not against them, and this requires making a genuine effort to understand such groups and their problems.

At the bottom of all these weaknesses is an elitist bias in a part of the animal rights movement centering around tourism, filmmaking and other such activities. While there is a legitimate limited role for all these activities, if necessary precautions are taken and these are linked to the wider objectives of reducing the sufferings of animals and maintaining ecological balance, serious problems arise when commercial considerations start dominating and real objectives are forgotten. The result is that while a handful of elites amuse themselves, ordinary villagers become increasingly alienated from otherwise highly justified movement for protecting the rights of other forms of life.

The task of protecting animals and birds has to be carried out with the help of ordinary people, not tourists. Therefore, it is important to include the important task of protection of animals in the wider frame work of social change in such a way that ordinary people see it as their moral duty to reduce this distress as much as possible. Once this attitude exists on a large scale, then countless ways of reducing the distress of other forms of life will be found by people themselves.

There is in fact a rich tradition in many communities respecting the rights of other forms of life, going in some cases to opposing any form of violence against them. Such traditions have been disrupted in some cases because of the reckless ways in which some mercenary interests have taken advantage of the poverty of people to pay them to collect frogs and butterflies or to hunt valuable species like musk deer.

Instead, why can not forest dwellers be given modest salaries to collect information on the mercenary elements who are destroying the botanical and zoological wealth of forests? Why can not they be mobilized to collect information on the various variety of birds and insects that exists in the forests after imparting them a little training for this work? A network of such trained barefoot zoologists can prove invaluable in knowing more about our wild life and in protecting it. In this and other ways, the conflicts that appear to exist between protection of animals and livelihood of people can be removed as much as possible and instead protection of animals can become the livelihood of many more people, particularly forests dwellers.

Case Studies of Harmonious Co-existence

1. Van Gujjars

Van Gujjars are a colourful community of nomadic and semi-nomadic pastoral people who live in the West Himalayan hills and the plains immediately below these hills. In the winter they live in forests of the plains with their buffaloes and in the summer most of them migrate to high altitude hills.

They are a very colourful people and in particular the long hand-woven topis several of them wear on their heads are very beautiful. They are Muslims by religion yet they have a predominantly vegetarian diet of mainly roti, lassi and other milk products. They live in remote forests far away from urban influence. They are very well informed about the flora and fauna of forests, including several medicinal plants. The indigenous breeds of buffaloes maintained by them for several generations are particularly suited to their way of life and can walk long distances on difficult hill paths without losing their way. Travellers and pilgrims to Garhwal, Kumaon and Himachal Pradesh often encounter these nomads without knowing that much of the milk supply for their use is provided by them.

These closely-knit people who are simple in their habits, graceful in their conversation and colourful in their appearance have been in the thick of controversy which is not at all of their own making. Instead, this controversy arises from the existing wildlife laws which say that human settlements and economic activities cannot be allowed in forest areas which have been declared as National Parks. This means that a large number of Van Gujjar (particularly those based in and around Dehra Dun district) have to give up their present settlements as well as the livelihood and pattern of life they have held for several generations. The prospects faced by many of them have further worsened because alternative sites they have been offered at a placed called Pathri are very unsatisfactory and not at all conducive to their present pattern of life. These sites were developed without any participative interaction with the Van Gujjar community.

A new beginning can certainly be made by a sympathetic inter-action with Van Gujjars to find out how many of them are in favour of a permanent resettlement site, and what should be the minimum area and other requirements of this site for them to earn an adequate livelihood. Experts also need to consider other problems of adjustments that people and their buffaloes used to a migratory way of life can face.

However, even if we assume that a satisfactory resettlement plan can be prepared on paper, it is unlikely that good quality resettlement land will be available in this region. As the experience of resettling evictees of Tehri dam has shown, it has been very difficult to find land for all of them even after trees were cut on a large tract of land. Another wave of displaced people is expected after the construction of other planned dams in this region.

Taking into consideration all these factors, it is difficult to dispute the view that any further displacement in this region should be avoided as far as possible. In the case of Van Gujjars, this displacement is certainly avoidable as they have lived in harmony with wild life for a long time and can continue to do so in future. All that is needed is a change in law allowing those communities to remain in National Park areas about whom it can be said with confidence that they will assist wild life protection instead of hindering it.

A Dehra Dun based voluntary organisation called Rural Litigation & Entitlement Kendra (RLEK) has actually prepared a plan for involving Van Gujjars in the protection of forests and wild life. RLEK Chairperson Avdhash Kaushal says with confidence, "We are quite sure that wild life will prosper if Van Gujjars are allowed to manage a part of the park on an experimental basis. If this plan is not successful, then the management can be taken back from them, but we are quite sure that this will not be necessary".

If they are adequately supported by the government then Van Gujjars can be very useful in taking effective action against poachers. They actually live in forests all the time and are in a good position to not only keep a watchful eye but also to challenge the poachers and curb their activities.

In addition, if some minimum training support is provided then Van Gujjars can prove invaluable in collecting information on fauna and flora, particularly medicinal plans. Some ayurvedic vaids of the region already acknowledge their debt to the Van Gujjars.

A base for such work of preparing barefoot zoologists and botanists among Van Gujjars has been already created as the RLEK has implemented an Innovative Literacy programme among the Van Gujjars during the last years. Very interesting literacy primers were prepared and some teachers accompanied Van Gujjars all the way to the high attitudes. On the other hand some teachers were also recruited from the hills and they accompanied Van Gujjars to the plains.

In addition RLEK has implemented a health programme, a veterinary care programme and a scheme for the direct sale of their milk in Dehradun. For the first time their names have been enlisted on the voters list and they have not only voted in the panchayat elections but also contested them. All this has helped to create the conditions of mutual trust in which the further training of Van Gujjars as barefoot botanists and zoologists can be done effectively. If a sincere effort is made in this direction, then this can become one of the most innovative efforts with several possibilities of learning within it and learning from it.

Camp Fire

According to a recent report by Stephen Kasere titled 'Campfire: Creating peace between man and beasts' a somewhat similar approach has given encouraging results in Zimbabwe. The introduction to this report say, "Campfire is a rural development programme in Zimbabwe, which aims at restoring the management of wildlife and other resources to the local inhabitants. It is changing people's attitude - they no longer view wild animals as pests that should be eradicated but as economic resources to be guarded for posterity. Campfire has spread from two districts in 1989, when programme began, to 23, enabling Zimbabwe's rural poor to reap the profits gained from using their natural resources sustainably."

Rara Avis

Similarly another report by Olivier Van Bogaert titled 'The Butterflies of Rara Avis' (Costa Rica)' says, "Rara Avis, a privately-owned nature reserve on the eastern edge of Braulio Carrillo National Park in Costa Rica, is demonstrating that it is possible to have ecological-sound activities which generate profits without harming the primary rain forests.

In addition to ecotourism and seed harvesting, the reserve has a butterfly farm which both conducts research and produces pupae for export. In this way, the project shows neighbouring farmers how a concrete alternative to cutting down forests can bring them both satisfaction and profits.

Vedda Tribals

There is another touching story from Sri Lanka which indicates that wisdom ultimately lies in working with the people instead of dislocating them. A large number of Vedda tribals were displaced some years back to make way for the Madura Oya National Park. However, a small number of Vedda tribals stayed back with their venerable chief and continued to struggle for the return of Veddas to their home. Gradually as deforestation continued at a brisk pace some senior officials started having second thoughts about the contribution the Veddas could have made to the protection of forests with their detailed knowledge and firm roots in the area. The late President Premdasa also took initiative in forming a trust to protect their culture. Finally the government gave permission to a group of Vedda to return to a 600-hectare tract in the Madura Oya Park and run it as a park Sanctuary. Volunteers from a local NGO the Cultural Survival Trust of Sri Lanka are working with the Vedda to draw up a Park Management Plan (This story of homecoming has been told by Patrik Harrigan and Mallika Wanigasundara recently in a Panos publication.)

Home coming is always good - even if this is after a long time and after a lot of injustice, but it'll be even better if the injustice of needless eviction can be avoided in the first place. People need not be displaced from parks, instead we should work with them to create better parks.

Appendix III

(A). Detailed Description of Field Visits with Van Gujjars

Date	Range/Area	Block	No. Deras Visited	No. Adult Interviewed	Overnight?
15/8/95	Gohri	Kunnaun	1	8	N
16/8/95	Laldhang	Jatpura	1	4	N
26/8/95	Mori Hills	--------	3	6	Y
27/8/95	Mori Hills	--------	1	3	Y
28/8/95	Mori Hills	--------	1	5	Y
29/8/95	Mori Hills	--------	1	5	Y
20/9/95	Yamuna Brdg	--------	1	4	N
30/9/95	Yamuna Brdg	--------	2	8	Y
26/10/96	Shahkumbar	Shahkumbar	3	10	Y
27/10/95	Shahkumbar	Sahsara	3	6	Y
31/10/95	Mohand	Mohand	2	6	Y
31/10/95	Mohand	Khajnawar	3	8	Y
1/11/95	Mohand	Khajnawar	3	9	Y
2/11/95	Mohand	Khajnawar	3	8	Y
13/11/95	Chillawali	Andheri	2	4	N
14/11/95	Dholkhand	Bam	2	3	Y
27/11/95	Pathari	Pathari	4	12	N
27/11/95	Gohri	Kunnaun	1	2	Y
30/11/95	Mohand	Kaluwala	1	3	Y
31/11/95	Mohand	Kaluwala	1	2	N
13/12/95	Kansrao	Bahera.	1	2	Y
14/12/95	Kansrao	Bahera.	1	8	N
14/12/95	Kansrao	Katapathar	1	2	Y
15/12/95	Kansrao	Katapathar	1	2	N
15/12/95	Kansrao	Pilkhan Sot	1	3	N
15/12/95	Ramgarh	Mohamedpur	1	10	Y
16/12/95	Ramgarh	Mohamedpur	1	2	N
16/12/95	Dholkhand	Rasulpur	1	3	Y
17/12/95	Dholkhand	Bam	1	8	N
17/12/95	Dholkhand	Gholna	1	10	Y
18/12/95	Ranipur	Ranipur	1	1	N
8/1/96	Ramgarh	Asarori	2	2	N
8/1/96	Ramgarh	Phandowala	1	2	Y
9/1/96	Ramgarh	Balandowal	2	5	N
9/1/96	Motichur	Koelpura	1	3	Y
10/1/96	Motichur	Colony	1	1	N
19/1/96	Laldhang	Sigaddi	1	100	Y
20/1/96	Chilla	Luni	4	10	Y
29/1/96	Ranipur	Ranipur	2	2	N

29/1/96	Ranipur	Rawli	1	20	Y
30/1/96	Ranipur	Chirak	1	2	N
30/1/96	Ranipur	Harnaul	2	10	N
30/1/96	Chillawali	Andheri	1	2	Y
31/1/96	Dholkhand	Beribada	1	2	N
2/2/96	Kansrao	Aamsot	3	4	N
2/2/96	Kansrao	Danda	2	2	N
2/2/96	Kansrao	Bahera	1	5	Y
11/2/96	Motichur	Koelpura	2	4	N
13/2/96	Mohand	Khajnawar	1	2	N
16/2/96	Chillawali	Chillawali	2	3	Y
17/2/96	Chillawali	Baniawala	1	2	N
17/2/96	Chillawali	Sukh	2	2	N

(B). Meetings with Van Gujjars

Date	Location	No. Adult Gujjars
30/11/95	Kaluwala	15
5/12/95	RLEK Office	15
16/1/96	Mohand	14
19/1/96	Laldhang/Chilla	12
23/1/96	Banbaya	5

(C). Meetings/Discussions with Local Villagers

Date	Location	Persons
30/8/95	Sankri Village	3
31/8/95	Mori Village	2
7/12/95	Buggawala Village	6
10/12/95	Rasulpur Village	2
11/12/95	Rasulpur Village	6
11/12/95	Elephant Corridor	2
18/12/95	Elephant Corridor	1
23/1/96	Haripur Taungya	15
19/2/96	Khushalipur	5
20/2/96	Shahjahanpur Paleo	4
20/2/96	Kaluwala Taungya	6
20/2/96	Jahanpur	4
20/2/96	Ganeshpur	2

(D). Consultation with Stakeholders

Date	Organization
9/11/95	Wildlife Institute of India
4/12/95	DISHA/Saharanpur
26/12/95	Centre for Science & Envir.
26/12/95	WWF-India
26/12/95	Centre for Envir. Law-WWF
27/12/95	SPWD/Delhi
16/1/96	VIKALP/Saharanpur
18/1/96	Friends of Doon
23/1/96	Wildlife Institute of India
26/1/96	Conservator, Meerut Circle, Forest Dept.
27/1/96	Himalayan Action Research Centre
7/2/96	Conservator, Meerut Circle with 7 DFOs
13/2/96	Chillawali Range Forest Officials
19/2/96	Mohand Range Forest Guard

Appendix IV

(A). Agenda for the National Consultation workshop on February 24-25, 1996.

Workshop on Community Forest Management of Protected Areas :
Van Gujjar Proposals for the Rajaji Park Area

24th February, 1996

Inaugural Function	11:30a.m.	Presided by Mr. D. Bandhopadhyay Chief Guest Mr. T.S.R.Prasada Rao
LUNCH	1:30 p.m.	
1st Session	2:00 p.m.	Van Gujjar Cultural Traits by Dr. Edward Glanville, Dr. Pernille Gooch and Van Gujjars Presided by Mr. Ravi Sharma
2nd Session	3:30 p.m.	CFM-PA Plan by Dr. Alan Warner Presided by Ms. Suneeta Narain

25th February 1996

1st Session	9:30a.m.	Legal Validity of Proposed Plan by Delhi University Law Department Presided by Prof. M.P. Singh
2nd Session	10:30a.m.	Training Module for Community Management Participation by Mr. S.Chandola Presided by Mr. Shyam Lal
3rd Session	11:30a.m.	Legal Remedies for Long Term Management Plans by Prof. M.P. Ramesh
4th Session	1:00 p.m.	Roundup by Mr. Avdhash Kaushal
Lunch	1:30 p.m.	

(B) List of participants for the National Consultation workshop

Van Gujjar And Local Village Participants

Mohammed Alam,
Chilla Ponurhole,
District Pauri Garhwal.

Sain Ali,
Asarori,
Ramgarh Range.
District Dehra Dun.

Mr. Munni Lal,
Taungya Pradhan,
Haripur Taungya
P.O. Buggawala,
District Haridwar

Dhumman Pradhan,
Beribada
Rajaji Park,
District Haridwar.

Noor Alam,
Rajaji Park
District Dehra Dun.

Alfa,
Mohand,
District Saharanpur.

Taalib Pradhan,
Kaluwala Khol
District Saharanpur.

Mustooq Lambardar,
Gohri Range, Rajaji Park,
District Pauri Garhwal.

Firozdin Pradhan,
District Dehra Dun.

Noor Ahmed,
Rajaji Park,
District Dehra Dun

Veeran
Khajnawar Khol
District Saharanpur

Sen Bibi
Mohand Khol
District Saharanpur

Alia ,
Chillawali range
District Haridwar

Kasim Pradhan,
District Dehradun

Shankar Dutt Upreti
Village Rasulpur
Distt Pauri Garhwal

Regional Participants

Dr. T. S. R. Prasada Rao,
Director, Indian Institute of Petroleum,
Mohkampur, Dehra Dun.

Mr. Rajesh Kumar,
People's Science Institute,
252, Vasant Vihar, Phase-I,
Dehra Dun.

Dr. Indira Kholi,
Social Scientist,
36, Balbir Road,
Dehra Dun.

Prof P. Dangwal,
D.A.V.P.G. College,
15-2/1, Lytton Road,
Dehradun.

Mr. Shyamlal, IFS,
Chief Conservator of Forest UP,
Bareilly.

Dr. Sanjay Chaudhary,
Veterinary Doctor,
M-16 Chanderlok Colony,
Rajpur Road,
Dehradun.

Ms. Shashi Sharma,
P.T.I (Bhasha),
2 Laltaro Pur Railway Road,
Haridwar

Mr. Ganga Sharan Sharma,
Kalyug Darpan,
Brahma Keerti,
Kali Mandir,
Bhungoda,
Haridwar.

Mr. Ashok P. Mishra,
Himachal Times,
Rajpur Road, Dehra Dun.

Dr. Karen Trollope Kumar
MBBS DCH
233-Vasant Vihar, Phase II,
Dehra Dun.

Mr. Bhupender Pal Singh,
Ramanand Cottage
Charlivile Road,
Mussoorie 248 179

Mr. Debashish Sen,
People's Science Institute,
252, Vasant Vihar, Phase-I,
Dehra Dun.

Mr. S. Chandola, IFS,
Conservator of Forests
Meerut Circle
Meerut.

Mr. C.P. Goyal, IFS
Divisional Forest Officer
Meerut.

Mr. S.K.Upadhyay,
M-16 Chanderlok Colony,
Rajpur Road,
Dehra Dun.

Mr. D.V.Singh,
S.D.O. Forests,
Shiwalik Forest Division,
5 Tilak Road,
Dehradun.

Mr. K.M.Rao, IFS
Divisional Forest Officer,
Shiwalik Forest Division,
6 Tilak Road,
Dehradun.

Ms. Bindu Kalia,
151/32 Jakhan,
Rajpur Road,
Dehradun 248001

Mr. Joseph Antony
26, Society Enclave,
Clement Town,
Dehradun

Mr. Praveen Kaushal
21 E.C.Road,
Dehradun.248001

Mr. Chatar Singh
Laxmanpur
P. O. Vikasnagar
Dehra Dun

Ms. Geetanjali Daundhiyal
145, Dharampur,
Dehra Dun

Ms. Preeti Dhobal
14-E Old Survey Road,
Dehra Dun

Dr. Motiya Sharan
M-16 Chanderlok Colony,
Rajpur Road,
Dehradun

National Participants

Ms. Anupama Sharma,
Faculty of Law,
University of Delhi
E-3/16, Janakpuri,
New Delhi.

Ms. Sunita Narain,
Deputy Director
Centre for Science and
Environment
41, Tuglakabad,
Institutional Area,
New Delhi-110062.

Ms. Sushma Sharma,
Faculty of Law,
Delhi,
Campus Law Centre.
New Delhi.

Ms. Neena Singh,
Centre for Science and
Environment,
41, Tuglakabad,
Institutional Area,
New Delhi-110062.

Prof. M.P. Singh,
Head, Faculty of Law,
University of Delhi,
Delhi 110007.

Prof. M.K. Ramesh,
Senior Asst. Professor,
National Law School of
India University,
Bangalore 560072.

Mr. Radhakrishna Rao,
Science Writer & Editorial Consultant,
1921, 5th Cross, IInd Phase,
J.P. Nagar, Bangalore.

Ms. Shalini Singh,
Faculty of Law,
University of Delhi,
Delhi 110007.

Prof. B.B. Pandey,
Faculty of Law,
University of Delhi,
Delhi.

Mr. Ashok Aggarwal,
Supreme Court Advocate
66 Babar Road,
New Delhi.

Dr. Ram Babu,
Deputy Director,
National Environment Engineering
Institue, (NEERI)
Nehru Marg, Nagpur

Ms. Nitya Ramakrishnan,
Supreme Court Advocate,
66 Babar Road,
New Delhi 110001,
Tel 3714531.

Dr. S. Maudgal,
Senior Advisor
Ministry of Environment, Forests
and Wild Life,
Government of India,
New Delhi - 110003

Mr. C.S.G. Roto,
National Environment Engineering
Institute, (NEERI)
Nehru Marh, Nagpur - 440020

Mr. C.K. Chandramohan,
The Hindu,
57/1, Scheme-7,
Shastri Nagar, Meerut

Mr. Simron Jit Singh,
D-917, Netaji Nagar
New Delhi - 110023

Mr. K.K. Sharma
Senior Deputy Director
LBSNNA, Charelvile, Mussoorie

Mr. Shan R. Haider
LBSNAA
Mussoorie,

Mr. Anil Maheshwari,
The Hindustan Times,
NewDelhi

Mr. V.J. Thomas,
Chief Sub Editor,
The Times of India,
7, Bahadur, Zafar Marg,
New Delhi - 110001

Mr. Jagdish Bhatt,
Principal Correspondent
The Times of India
Flat No 5, Type IV,
Richmond, Shimla

Mr. D. Bandyopadhyay,
Chairman Indian, Institute
of Management,
58-C, Block-D, New Alipore
Calcutta - 700053

International Participants

Dr. Pernille Gooch,
Professor, Sociology Department,
Lund University,
Sweden.

Ms. Ginny Point
Nutritionist
RR # 3 Site 1A
Box 1 Comp 10
Centreville, Nova Scotia
BOP 1JO
Canada

Dr. Alan Warner,
Professor, School of Education
Acadia University
Wolville, Nova Scotia
Canada.

Dr. Edward Glanville,
Department of Anthropology
McMaster University,
Hamilton Ontario,
Canada

Appendix - V

DESCRIPTION OF METHODOLOGY FOR RLEK VAN GUJJAR CENSUS

Rural Litigation Entitlement Kendra (RLEK) initiated a census of the numbers of Van Gujjars and Van Gujjar animals in the area of the proposed Rajaji Park in November 1992. Previous records of the number of Van Gujjars have been grossly inaccurate, as the last official census dates back to 1937 and is hopelessly outdated while more recent informal efforts have faced major obstacles and threats to their validity. In particular, because of the Van Gujjars distrust of local authorities, they are not quick to volunteer accurate responses about sensitive information, that they feel might in future affect government policies toward them. Moreover, they feel that surveys could be used by the government in an attempt to remove them from the proposed park area. As the result, before initiating a detailed survey, it was essential for RLEK to overcome their suspicions.

First, the four key RLEK staff, who co-ordinated the survey and its implementation, had worked extensively with the Van Gujjars over the past year and were well known to a wide range of Van Gujjars as advocates for them. In particular, RLEK had been instrumental in helping the Van Gujjars resist the attempt by the Forest Department to prevent them from re-entering the proposed park area at the end of their another migration in 1992. Second, the information from the survey was to be used to plan the adult literacy program for the Van Gujjars, who were most enthusiastic about the literacy program, and hence saw the survey as a positive in planning it.

The census was initiated through a meeting to define the required information: number of people according to gender, age, nomadic status, and clan; and number of buffalo. Eighteen staff were then selected to define the process and divide up the area. which covered 25 ranges and one district.

In some instances one person covered only one range while others took on several ranges.

The survey procedure was as follows: The surveyor visited several deras in the range to determine the location of all deras by using a map of the range. Information from several deras was cross-checked in order to establish the accuracy of the map. The surveyor then proceeded to visit each dera identified on the map to collect information. During the initial phase of the survey from November 1992 to February 1993 there were large monthly meetings which served as a double check on the throughout comprehensiveness of the survey. These meetings were called to discuss immediate Van Gujjar problems, and included representatives from most deras. At the December meeting, a number of representatives voiced their concern that their deras had not been surveyed as yet; they were worried this would prevent them from participating in the literacy program. The surveyors noted the deras missed (some had simply not been reached as yet). At the subsequent January meeting, there were but a few complaints, and very few thereafter. The, last few deras were identified in November 1993, and were added to the survey during the final implementation phase of the literacy program. These deras were missed as they had changed locations in 1992 and had not been identified in their new areas.

The surveyor was usually expected, and warmly greeted upon arrival ("the counting man has come"). The information was collected through informal discussion in line with the social customs of the Van Gujjars and was immediately recorded in a notebook. The number of buffalo held by each dera was the most sensitive piece of information as the Van Gujjars would tend to under report their holding of buffalo. As the result, this information was collected by visual counting, as buffalo remain near the dera during the day. In addition; the surveyor would ask about the number of buffalo at a given dera when visiting other deras in the area, as a cross-check.

THE RIO DECLARATION ON ENVIRONMENT AND DEVELOPMENT

The Declaration is reproduced in full below.

The United Nations Conference on Environment and Development.

Having met at Rio de Janerio from 3 to 14 June 1992

Reaffirming the Declaration of the United Nations Conference on the Human Environment, adopted at Stockholm on 16 June 1972, and seeking to build upon it,

With the goal of establishing a new and equitable global partnership through the creation of new levels of cooperation among States, key sectors of societies and people,

Working towards international agreements which respect the interests of all and protect the integrity of the global environmental and development system,

Recognizing the integral and interdependent nature of the Earth, our home,

Proclaims that:

Principle 1 Human beings are at the centre of concerns for sustainable development. They are entitled to a healthy and productive life in harmony with nature.

Principle 2. States have, in accordance with the Charter of the United Nations and the principles of international law, the sovereign right to exploit their own resources pursuant to their own environmental and developmental policies, and the responsibility to ensure that activities within their jurisdiction or control do not cause damage to the environment of other States or of areas beyond the limits of national jurisdiction.

Principle 3. The right to development must be fulfilled so as to equitably meet developmental and environmental needs of present and future generations.

Principle 4. In order to achieve sustainable development, environmental protection shall constitute an integral part of the development process and cannot be considered in isolation from it.

Principle 5. All States and all people shall cooperate in the essential task of eradicating poverty as an indispensible requirement for sustainable development, in order to decrease the disparities in standards of living and better meet the needs of the majority of the people of the world.

Principle 6. The special situation and needs of developing countries, particularly the least developed and those most environmentally vulnerable, shall be given special priority. International actions in the field of environment and development should also address the interests and needs of all countries.

Principle 7. States shall cooperate in a spirit of global partnership to conserve, protect and restore the health and integrity of the Earth's ecosystem. In view of the different contributions to global environmental degradation, States have common but differentiated responsibilities. The developed countries acknowledge the responsibility that they bear in the international pursuit of sustainable development in view of the pressures their societies place on the global environment and of the technologies and financial resources they command.

Principle 8. To achieve sus.. .ui development and a higher quality of life for all people, States should reduce and eliminate unsustainable patterns of production and consumption and promote appropriate demographic policies.

Principle 9. States should cooperate to strengthen endogenous capacity building for sustainable development by improving scientific understanding through exchanges of scientific and technological knowledge, and by enhancing the development, adaptation, diffusion and transfer of technologies, including, new and innovative technologies.

Principle 10. Environmental issues are best handled with the participation of all concerned citizens, at the relevant level. At the national level, each individual shall have appropriate access to information concerning the environment that is held by public authorities, including information on hazardous materials and activities in their communities, and the opportunity to participate in decision making process. States shall facilitate and encourage public awareness and participation by making information widely available. Effective access to judicial and administrative proceedings, including redress and remedy, shall be provided.

Principle 11. States shall enact effective environmental legislation. Environmental standards, management objectives and priorities should reflect the environmental and developmental context to which they apply. Standards applied by some countries may be inappropriate and of unwarrant economic and social cost to other countries, in particular developing countries.

Principle 12. States should cooperate to promote a supportive and open international economic system that would lead to economic growth and sustainable development in all countries, to better address the problems environmental degradation. Trade policy measures for environment crimination or a disguised restriction on international trade. Unilateral actions to deal with environmental challenges outside the jurisdiction of the importing country should be avoided. Environmental measures addressing transboundary or global environmental problems should, as far as possible be based on an international consensus.

Principle 13. States shall develop national law regarding liability and compensation for the victims of pollution and other environmental damage States shall also cooperate in an expeditious and more determined manage to develop further international law regarding liability and compensation for adverse effects of environmental damage caused by activities within their jurisdiction or control to areas beyond their jurisdiction.

Principle 14. States should effectively co-operate to discourage or prevent the relocation and transfer to other States of any activities and substances that cause severe environmental degradation or are found to be harmful to human health.

Principle 15. In order to protect the environment, the precautionary approach shall be widely applied by States according to their capabilities. Where there are threats of serious or irreversible damage, lack of full scientific certainity shall not be used as a reason for postponing cost effective measures to prevent environmental degradation.

Principle 16. National authorities should endeavour to promote the internalization of environmental costs and the use of economic instruments, taking into account the approach that the polluter should, in principle, bear the cost of pollution, with due regard to the public interest and without distorting international trade and investment.

Principle 17. Environmental impact assessment, as a national instrument, shall be undertaken for proposed activities that are likely to have a significant adverse impact on the environment and are subject to a decision of a competent national authority.

Principle 18. States shall immediately notify other States of any natural disasters or other emergencies that are likely to produce sudden harmful effects on the environment of those States. Every effort shall be made by the international community to help States so afflicted.

Principle 19. States shall provide prior and timely notification and relevant information to potentially affected States on activities that may have a significant adverse transboundary environmental effect and shall consult with those States at an early stage and in good faith.

Principle 21. The creativity, ideals and courage of the youth of the world should be mobilized to forge a global partnership in order to achieve sustainable development and ensure a better future for all.

Principle 22. Indigenous people and their communities, and other local communities, have a vital role in environmental management and development because of their knowledge and traditional practices. States should recognize and duly support their identity, culture and interests and enable their effective participation in the achievement of sustainable development.

Principle 23. The environment and natural resources of people under oppression, domination and occupation shall be protected.

Principle 24. Warfare is inherently destructive of sustainable development. States shall therefore respect international law providing protection for the environment in times of armed conflict and cooperate in its further development, as necessary.

Principle 25. Peace, development and environmental protection are interdepend and indivisible.

Principle 26. States shall resolve all their environmental disputes peacefully and by appropriate means in accordance with the Charter of the United Nations.

Principle 27. States and people shall cooperate in good faith and in a spirit of partnership in the fulfillment of the principles embodied in this Declaration ad in the further development of international law in the field of sustainable development.

Appendix - VII

NON-LEGALLY BINDING AUTHORITATIVE STATEMENT OF PRINCIPLES FOR A GLOBAL CONSENSUS ON THE MANAGEMENT, CONSERVATION AND SUSTAINABLE DEVELOPMENT OF ALL TYPES OF FORESTS

PREAMBLE

(a) The subject of forests is related to the entire range of environmental and development issues and opportunities, including the right to socio-economic development on a sustainable basis.

(b) The guiding objective of these principles is to contribute to the management, conservation and sustainable development of forests and to provide for their multiplying and complementary conditions and now.

(c) **Forestry issues and opportunities should be examined in a holistic and balanced manner within the over all context of environment and development, taking into consideration the multiple functions and uses of forests, including traditional uses are constrained or restricted, as well as the potential for development that sustainable forest management can offer.**

(d) These principles reflect a first global consensus on forests. In committing themselves to the prompt implementation of these principles, countries also decide to keep them under assessment for their adequacy with regard to further international cooperation on forest issues.

(e) These principles should apply to all types of forests, both natural and planted, in all geographical regions and climatic one including natural, boreal, sub-temperate, temperate, subtropical and tropical.

(f) All types of forests body compelled and unique ecological processes which are the basis for their present and potential capacity to provide resources to satisfy human as well as environmental values, and as such their sound management and conservation is of concern to the Governments of the countries to which they belong and are of value to local communities and to the environment as a whole.

(g) Forests are essential to economic development and the maintenance of all forms of life.

(h) Recognizing that the responsibility for forest management, conservation and sustainable development is in many states allocated among federal/national,state/provincial and local levels of government, each State, in accordance with its constitution and/or national legislation, should pursue these principles at the appropriate level of government.

PRINCIPLES / ELEMENTS

1. (a) States have, in accordance with the Charter of the United Nations and the principles of international law, the sovereign right to exploit their own resources pursuant to their own environmental policies and have the responsibility to ensure that activities within their jurisdiction or control do not cause damage to the environment of other states or of areas beyond the limits of national jurisdiction.

(b) The agree full incremental cost of achieving benefits associated with forest conservation and sustainable development requires increased international cooperation and should be equitable and shared by the international community.

2. (a) States have the sovereign and inalienable right to utilize, manage and develop their forests in accordance with their development needs and level of socio-economic development and on the basis of national policies consistent with sustainable development and legislation, including the conversion of such areas for other uses within the overall socio-economic development plan and based on national land use policies.

(b) Forest resources and forest land should be sustainable managed to meet the social, economic, ecological, cultural and spiritual needs of present and future generations. These roads are for forest products and services, such as wood and wood products, water, food fodder, medicine, fuel, shelter, employment, recreation, habitats for wildlife, landscape diversity, carbon sinks and reservoirs, and for other forest products. appropriate measures should be taken to protect forests against harmful effects of pollution, including air-borne pollution, fires, pasts and diseases, in order to maintain their full multiple value.

(c) The provision of timely, reliable and accurate information of forests and forest ecosystems is essential for public understanding and informed decision making and should be ensured.

(d) Government should promote and provide opportunities for the participation of interested parties, including local communities and indigenous people, industries, labour, non governmental organisations and individual forest dwellers and women in the development, implementation planning of national forest policies.

3. (a) National policies and strategies should provide a framework for increased efforts, including the development and strengthening of institutions and programmes for the management, conservation and sustainable development of forests and forest lands.

(b) International institutional arrangements, building on those organisations and mechanism already in existence, as appropriate, should facilitate international cooperation in the field of forest.

(c) All aspects of environmental protection and social and economic development as they relate to forests and forest lands should be integrated and comprehensive.

4. The vital role of all types of forest in maintaining the sociological processes and balance at the local, national, regional and global levels through, inter alia, their role in protecting fragile ecosystems, watersheds and fresh water resources and as rich storehouses of biodiversity and biological resources and sources of genetic material for biotechnology products, as well as photosynthesis, should be recognised.

5. (a) National forest policies should recognize and duly support the identity, culture and the rights of the indigenous people, their communities and other communities and forest dwellers. Appropriate conditions should be promoted for these groups to enable them to have an economic stage in forest use, perform economic activities, and achieve and maintain cultural identity and social organisation, as well as adequate levels of livelihood and well being, through, inter-alia, those land tenure arrangements which serve as incentives for the sustainable management of forests.

(b) The full participation of women is all aspects of the management, conservation and sustainable development of forests should be actively promoted.

6. (a) All types of forests play an important role in meeting energy requirements through the provision of a renewable source of bio-energy, particularly in developing countries, and the demands for fuel wood households and industrial needs should be met through sustainable forest management, afforestation and reforestation. To this end, the potential contribution of the plantation of indigenous and introduced species for the provision of both fuel and industrial wood should be recognised.

(b) National policies and programmes should take into account the relationship, where it exists, between the conservation, management and sustainable development of forests and all aspects related to the production, consumption, recycling and/or final disposal of forest products.

(c) Decisions taken on the management, conservation and sustainable development of forest resources should benefit, to the extent practicable, from a comprehensive assessment of economic and non-economic values of forest goods and services and of the environmental costs and benefits. The development and improvement of mythologies for such evaluations should be promoted.

(d) The role of planted forests and permanent agricultural crops as sustainable and environmentally sound sources of renewable energy and industrial raw material should be recognized, enhanced and promoted. Their contribution to the maintenance of ecological processes, to offsetting pressure on primary/old growth forest and to providing regional employment and development with the adequate involvement of local inhabitants should be recognised and enhanced.

(e) Natural forests also constitute a source of goods and services, and their conservation, sustainable management and use should be promoted.

7. (a) Efforts should be made to promote a supportive international economic climate conductive to sustained and environmentally sound development of forests in all countries, which include, inter alia, the promotion of sustainable patterns of production and consumption, the eradication of poverty and the promotion of food security.

(b) Specific financial resources should be provided to developing countries with significant forest areas which establish programmes for the conservation of forests including protected natural forest areas. These resources should be directed notably to economic sectors which would estimate economic and social substitution activities.

8. (a) Efforts should be undertaken towards the up greening of the world. All countries, notably developed countries, should take positive and transparent action towards reforestation, afforestation and forest conservation, as appropriate.

(b) Efforts to maintain and increase forest cover and forest productivity should be undertaken in ecologically, economically and socially sound ways through the rehabilitation, reforestation and reestablishment of trees and forests on unproductive, degraded and deforested lands, as well as through the management of existing forest resources.

(c) The implementation of national polices and programmes aimed at forest management, conservation and sustainable development, particularly in developing countries, should be supported by international financial and technical cooperation, including through the private sector, where appropriate.

(d) Sustainable forest management and use should be carried out in accordance with national development policies and priorities and on the basis of environmentally sound national guidelines. In the formulation of such guidelines, account should be taken, as appropriate and if applicable, of relevant internationally agreed methodologies and criteria.

(e) Forest management should be integrated with management of adjacent areas so as to maintain ecological balance and sustainable productivity.

(f) National policies and/or legislation aimed at management, conservation and sustainable development of forests should include the protection of ecologically viable representative or unique examples of forests, including primary/old growth forests, cultural, spiritual, historical, religious and other unique and valued forests of national importance.

(g) Access to biological resources, including genetic material, shall be with due regard to the sovereign rights of the countries where the forests are located and to the sharing on mutually agreed term of technology and profits from biotechnology products that are derived from these resources.

(h) National policies should ensure that environmental impact assessments should be carried out where actions are likely to have significant adverse impacts on important forest resources, and where such actions are subject to a decision of a competent national authority

9. (a) The efforts of developing countries to strengthen the management, conservation and sustainable development of their forest resources should be supported by the international community, taking into account the importance of redressing external indebtedness, particularly where aggravated by the net transfer of resources to developed countries, as well as the problem of achieving at least the replacement value of forests through improved market access for forest products, especially processed products. In this respect, special attention should also be given to the countries undergoing the process of transition to market economies.

(b) The problems that hinder efforts to attain the conservation and sustainable use of forest resources and that item from the lack of alternative options available to local communities, in particular the urban poor and poor rural populations who are economically and socially dependent on forests and forest resources, should be addressed by Governments and the international community.

(c) National policy formulation with respect to all types of forests should take account of the pressures and demands imposed on forest ecosystems and resources from influencing factors out side the forest sector, and inter sectoral means of dealing with these pressure and demand should be sought.

10. Now and additional financial resources should be provided to developing countries to enable them to sustainably manage, conserve and develop their forest resources, including through affores-

tation, reforestation and combating deforestation and forest and land degradation.

11. In order to enable, in particular, developing countries to enhance their indigenous capacity and to better manage, conserve and develop their forests resources, the access to and transfer of environmentally sound technologies and corresponding know-how on favourable terms, including on concessional and preferential terms, as mutually agreed, in accordance with the relevant provisions of Agenda 21, should be promoted, facilitated and financed, as appropriate.

12.(a)Scientific research, forest inventories and assessments carried out by national institution which take into account, where relevant, biological, physical, social and economic variables, as well as technological development and its application in the field of sustainable forest management, conservation & development, should be strengthened through effective modalities, including international cooperation. In this context, attention should also be given to research and development of sustainably harvested non wood products.

(b) National and, where appropriate, regional and international economics, anthropology and social aspects of forests and forest management are essential to the conservation and sustainable development of forests and should be strengthened.

(c) International exchange of information on the results of forest and forest management research and development should be enhanced and broadened, as appropriate, making full use of education and training institutions, including those in the private sector.

(d) Appropriate indigenous capacity and local knowledge regarding the conservation and sustainable development of forests should, through institutional and financial support and in collaboration with the people in and, as appropriate, introduced in the implementation of programmes. Benefit arising from the utilization of indigenous knowledge should therefore be equitably shared with such people.

13. (a) Trade in forest products should be based on non-discriminatory and multilaterally agreed rules and procedures consistent with international trade law and practices. In this context, open and free international trade in forest products should be facilitated.

(b) Reduction or removal of tariff barriers and impediments to the provision of better market across and better prices for higher value added forest products and their local proceeding should be encouraged to unable producer countries to better conserve and manage their renewable forest resources.

(c) Incorporation of environmental costs and benefits into market forces and mechanisms, in order to achieve forest conservation and sustainable development, should be development, should be encouraged both domestically and internationally.

(d) Forest conservation and sustainable development policies should be internal economic trade and other relevant policies.

(e) Fiscal, trade, industrial, transportation and other policies and practices that may lead to forest degradation should be avoided. Adequate policies, aimed at management, conservation and sustainable development of forests, including, where appropriate, incentives, should be encouraged.

14. Unilateral measures, incompatible with international obligations or agreements, to restrict and/or ban international trade in timber or other forest products should be removed or avoided, in order to attain long term sustainable forest management.

15. Pollutants, particularly air-borne pollutants, including those responsible for acidic deposition, that are harmful to the health of forest ecosystems at the local, national, regional and global levels should be controlled.

Appendix - VIII

I L O Convention 169 of 1989

Convention Concerning Indigenous And Tribal Peoples In Independent Countries.

The General Conference of the International Labour Organisation,

Having been convened at Geneva by the Governing by the Governing Body of the International Labour office, and having met in its 76th Session on 7th June, 1989, and

Noting the International standards contained in the Indigenous and Tribal Populations Convention and recommendation, 1957, and Recalling the terms of the Universal Declaration of Human Rights, the International Convention on Economic, Social and Cultural Rights, the International Convention on Civil and Political Rights, and the many International instruments on the prevention of discrimination, and Considering that the developments which have taken place in international law since 1957, as well as developments in the situation of indigenous and tribal peoples in all regions of the world, have made it appropriate to adopt new international standards on the subject with a view to removing the assimilation orientation of the earlier standards, and

Recognising the aspirations of those people to exercise control over their own institution, ways of life economic development and to maintain and develop their identities, languages and religions, within the framework of the states in which they live, and

Noting that in many parts of the world these people are unable to enjoy their fundamental human rights to the same degree as the rest of the population of the states within which they live, and their laws, values, customs and perspective have often been eroded, and

Calling attention to the distinctive contributions of indigenous and tribal peoples to the cultural diversity of social and ecological harmony of human-kind and to international cooperation and understanding, and

Noting that the following provisions have been formed with the cooperation of the United Nations, the Food and Agriculture Organisation of the United Nations the Food and Agriculture organisations of the United Nations, the United Nations Educational, Scientific and cultural organisations and the World Health Organisation, as well as of the Inter American Indian Institute, at appropriate levels and in their respective fields, and that it is proposed to continue this co-operation in promoting and securing the application of these provisions, and

Having divided upon the option of certain proposals with regard to the partial revision of the Indigenous and Tribal Populations Convention, 1957 (No.107), which is the fourth item on the agenda of the session, and

Having determined that these proposals shall take to form of an international Convention revising the Indigenous and Tribal Population Convention, 1957; adopts this twenty-seventh day of June of the year one thousand nine hundred and eighty nine the following Conventions which may be cited as the Indigenous and Tribals People Convention, 1989:

PART I, CENTRAL POLICY

Article 1

1, This Convention applies to :
a) tribal peoples in independent countries whose social, cultural and economic conditions distinguish them from other sections of the national community, and whose status is regulated wholly or partially by their own customs or traditions or by special laws or regulations.

b) peoples in independent countries who are regarded as indigeous on account of their descent from the populations which inhabited the country, or a geographical region to which the country belongs, at the time of conquest or colonization or the establishment of present date boundaries and who, irrespective of their legal status, retain some or all of their own social, economic, cultural and political institutions.

2, Self identification as indigenous or tribal shall be regarded as a fundamental criterion for determining the grounds to which the provisions of this convention apply.

3, The use of the term "peoples" in this convention shall not be constructed as having any implications as regards the rights which may attach to the termunder international law.

Article 2

1, Governments shall have the responsibility for developing, with the participation of the peoples concerned, co-ordinated and systematic action to protect the rights of these peoples and to guarantee respect for their integrity.

2, Such action shall include measures for :

a) ensuring that members of these peoples benefit on an equal footing from the rights and opportunities which national laws and regulations grant to other members of the population;

b) promoting the full realization of the social, economic and cultural rights of these peoples with respect for their social and cultural identity, their customs and traditional and their institutions;

c) assisting the members of the peoples concerned to eliminate socio-economic gaps that may exist between indigenous and other members of the national community, in a manner compatible with their aspirations and ways of life.

Article 3

1, Indigenous and tribal peoples shall enjoy the full measure of human rights and fundamental freedoms without hindrance or discrimination. The provisions of the convention shall be applied without discrimination to male and female members of these peoples.

2, No form of force or coercion shall be used in violation of the human rights and fundamental freedoms of the peoples concerned, including the rights contained in this convention.

Article 4

1, Special measures shall be adopted as appropriate for safeguarding the persons, institutions, property, lobour, cultures and environment of the peoples concerned.

2, Such special measures shall not be contrary to the freely expressed wishes of the people concerned.

3, Enjoyment of the general rights of citizenship without discrimination, shall not be prejudiced in any way by such special measures.

Article 5

Un applying the provisions of this convention;

a) the social, cultural, religious and spiritual values and practices of these peoples shall be recognised and protected and due account shall be taken of the nature of the problems which face them both as groups and as individuals;

b) the integrity of the values, practices and institutions of these peoples shall be respected;

c) policies aimed at matching the difficulties experienced by these peoples in facing new conditions of life and work shall be adopted with the participation and cooperation of the peoples affected.

Article 6

1, In applying the provisions of this convention, governments shall:

a) consult the people concerned, through appropriate procedures and in particular through their representative institutions, whenever consideration is being given to legislative or administrative measures which may affect them directly;

b) establish means by which these peoples can freely participate to, to at least the same content as other sectors of the population, at all levels of decision making in elective institutions and administrative and other bodies responsible for policies and programmes which concern them;

c) establish means for the full development of these peoples' own institutions and initiatives, and in appropriate cases provide the resources necessary for this purpose.

2, The consultation carried cut in application of this Convention shall be undertaken, in good faith and in a form appropriate to the circumstances, with the objectives of achieving agreement or consent to the proposed measures.

Article 7

1, The peoples concerned shall have the **right to decide their own priorities for the process of development as it affects their lives, beliefs, institutions and spiritual well being and the lands they occupy or otherwise use, and to exercise control, to the extent possible, over their own economic, social and cultural development.** In addition they shall participate in the formulation, implementation and evaluation of plans and programmes for national and regional development which may effect them directly.

2, The improvement of the conditions of life and work and levels of health and education of the people concerned, with their participation and cooperation, shall be matter of priority in plans for the overall economic development of areas they inhabitat, special projects for developments of the areas in question shall also be so designed as to promote such improvement.

3, Governments shall insure that, whenever appropriate, studies are carried out, in co-operation with the peoples concerned, to assess the social, spiritual, cultural and environmental impact on them of planned development activities. The results of these studies shall be considered as fundamental criteria for the implementation of these activities.

4, Governments shall take measures, in-cooperation with the peoples concerned, to protect and reserve and conserve the environment of the territories they inhabitat.

Article 8

1, In applying national laws and regulations to the peoples concerned, due regard shall be had to their customs or customary laws.

2, These people shall have the right to retain their own customs and institutions, where these are not incompatible with fundamental rights defined by the national legal system and with internationally recognised human rights. Procedures shall be established, wherever necessary, to resolve conflicts which may arise in the application of this principle.

3, The application of paragraphs and 2 of this Article shall not prevent members of these peoples from exercising the rights granted to all citizens and from assuming the corresponding duties.

Article 9

1, To the extent compatible with the national legal system and internationally recognised human rights, the methods customarily practiced by the people concerned for dealing with offences committed by their members shall be respected.

2, The customs of these peoples in regard to penal matters shall be taken into consideration by the authorities and courts dealing with such cases.

Article 10

1, In imposing penalties laid down by general law on members of these peoples account shall be taken of their economic, social and cultural characteristics.

2, Preferences shall be given to methods of punishment other than confinement in prison.

Article 11

The exaction from members of the peoples concerned of compulsory personal services in any form, weather paid or unpaid, shall be prohibited and punishable by law, except in cases prescribed by law for all citizens.

Article 12

The peoples concerned shall be safeguarded against the abuse of their rights and shall be able to take legal proceedings, either individually or though their representative bodies, for the effective protection of these rights. Measures shall be taken to ensure that members of these peoples can understand and be understood in legal proceedings where necessary though the provision of interpretation or by other effective means,

PART II, LAND

Article 13

1, In applying the provisions of this part of the Convention governments shall respect the special importance for the cultures and spiritual values of the peoples concerned of their relationship with the lands or territories, or both as applicable, which they occupy or otherwise use, and in particular to collective aspects of the relationship.

2, The use of the term 'Lands' in Article 15 and 16 shall include the concept of territories, which covers the total environment of the areas which peoples concerned occupy or otherwise use.

Article 14

1, The rights of ownership and possession of the people concerned over the lands which they traditionally occupy shall be recognised. In addition, **measures shall be taken in appropriate cases to safeguard the right of the peoples concerned to use lands not exclusively occupied by them, but to which they have traditionally had access for their sustenance and traditional activities. Particular attention shall be paid to the situation of nomadic peoples and shifting cultivators in this respect.**

2, Governments shall take steps to necessary to identify the lands which the peoples concerned traditionally occupy, and to guarantee effective protection of their rights of ownership and possession.

Article 15

1, The right of the peoples concerned to the natural resources pertaining to their lands shall be specially safeguarded. these rights include the right of these peoples to participate in the use, management and conservation of these resources.

2, In cases in which the state retains the ownership of mineral or sub surface resources or rights to other resources pertaining to lands, governments shall establish or maintain procedures through which they shall consult these peoples, with a view to ascertaining weather and to what degree their interests would be prejudiced before under taking or permitting any programmes for the exploration or exploitation of such resources pertaining to their lands. The peoples concerned shall whenever possible participate in the benefits of such activities, and shall receive fair compensation for any damages which they may sustain as a result of such activities.

Article 16

1, Subject to the following paragraphs of this article, the peoples concerned shall not be removed from lands which they occupy.

2, Where the relocation of these people is considered necessary as an exceptional measure, such relocation shall take place only with their free and informed consent. Where their consent cannot be obtained, such relocation shall take place only following appropriation procedures established by national laws and regulations, including public inquires where appropriate, which provide the opportunity for effective representation of the people concerned.

3, Wherever possible, these people shall have the right to return to their traditional lands, as soon as the grounds for relocation ceases to exist.

4, When such return is not possible, as determined by agrement or, in the absence of such agreement, through approach procedures, these people shall be provided in all possible cases with lands of quality and legal ststus at least equal to that of the lands previously occupied by them, suitable to provide for their present needs and future development. Where the peoples concerned express a preference for compensation in money or in kind, they shall be so compensated under appopriate gurantees.

5, Persons thus relocated shall be fully compensated for any resulting loss or injury.

Article 17

1, Procedures established by the peoples concerned for the transmission of land rights among members of these peoples shall be respected.

2, The peoples concerned shall be consulted whnever consideration is being given to their capacity to alienate their lands or otherwise transmit their rights outside their own community.

3, Persons not belonging to these people shall be prevented from taking advantage of their customs or of lack of understanding of the laws on the part of their members to secure the ownership, possesion or use of land belonging to them.

Article 18

Adequate penalties shall be established by law for unauthorised intrusion upon, or use of, the lands of the peoples concerned, and governments shall take measures to prevent such offences.

Article 19

National agrarian programmes shall secure to the peoples concerned treatment equavilent to that accorded to other sectors of the population with regard to:

(a) the provision of more land for these peoples when they have not the area necessary for providing the essantials of a normal existance, or of any possible increase in their numbers;

(b) the provision of the means required to promote the development of the lands which these peoples already possess.

PART III, RECRUTIMENT AND CONDITIONS OF EMPLOYMENT

Article 20

1, Governments shall, within the framework of national laws and regulations, and un cooperation with the peoples concerned, adopt special measures to ensure the effective protection with regard to recrutiment and conditions of employment of workers belonging to these peoples, to the extent that they are not effectevly protected by laws applicable to workers in general.

2, Governments shall do every thing possible to prevent any descrimination between workers belonging to the peoples concerned and other workers, in particular as regards:

(a) admission to employment, including skilled employment, as well as measures for the promotion and advancment;

(b) equal renumeration for work of equal value;

(c) medical and social assistance, occupational safety and health, all social security benefits and any other occupationally related benefits, and housing;

(d) the right of association and freedom for all lawful trade union activities, and the right to conclude collective agreements with employees or employers organisations.

3, The measures taken shall include measures to ensure:

(a) that workers belonging to the peoples concerned including sesional, casual and migrant workers in agricultural and other employment, as well as those employed by labour contractors, enjoy the protection afforded by national law and practice to other such workers in the same sectors, and that they are fully informed of their rights under labour legislation and of the means of redress available to them;

(b) that workers belonging to these peoples are not subjected to working conditions hazardous to their health, in particular through exposure to pesticides and other toxic substances;

(c) that workers belonging to these peoples are not subjected to coercive recruitment systems, included bonded labour and other forms of debt servitude;

(d) that workers belonging to these peoples enjoy equal opportunities and equal treatment in employment for men and women, and protection from sexual harassment.

4, Particular attention shall be paid to the establishment of adequate labour inspection services in the areas where workers belonging to the peoples concerned undertake wage employment in order to ensure compliance with the provisions of this part of the Convention.

PART IV VOCATIONAL TRAINING, HANDICRAFTS AND RURAL INDUSTRIES

Article - 21
Members of the peoples concerned shall enjoy opportunities at least equal to those of the other citizens in respect of vocational training measures.

Article - 22
1, Measures shall be taken to promote the voluntary participation of members of the peoples concerned in vocational training programmes of general application.

2, Whenever existing programmes of vocational training of general application do not meet the special needs of the peoples concerned, governments shall with the participation of these peoples, ensure the provision of special training programmes and facilities.

3, Any special training programmes shall be based on the economic environment, social and cultural conditions and practical needs of the peoples concerned. Any studies made in this connection shall be carried out in cooperation with these peoples, who shall be consulted on the organisation with these peoples shall progressively assume responsibility for the organisation and operation of such special training programmes if they so decide.

Article 23
1, Handicrafts, rural and community based industries, and sustenance economy and traditional activities of the peoples concerned such as hunting, fishing, trapping and gathering, shall be recognised as important factors in the maintenance of their cultures and in their economic self-reliance and development. Governments shall, with

the participation of these peoples and whenever appropriate ensure that these activities are strengthened and promoted.

2, Upon the request of the peoples concerned, appropriate technical and financial assistance shall be provided wherever possible, taking into account the traditional technologies and cultural characteristics of these peoples, as well as the importance of sustainable and equitable development.

PART V, SOCIAL SECURITY AND HEALTH

Article - 24

Social security schemes shall be extended progressively to cover the peoples concerned, and supplied without discrimination against them.

Article - 25

1, Governments shall ensure that adequate health services are made available to the peoples concerned, or shall provide them with resources to allow them to design and deliver such services under their own responsibility and control, so that they may enjoy the highest attainable standard of physical and mental health.

2, Health services shall, to the extent possible, be community- based. these services shall be planned and administrated in cooperation with the peoples concerned and take into account their economic, geographic, social and cultural conditions as well as their traditional preventive care, healing practices and medicines.

3, The health care system shall give preference to the training and employment of the local community health workers, and focus on primary health care while maintaining strong links with other levels of health care services.

4, The provision of such health services shall be co-ordinated with either social, economic and cultural measures in the country.

PART VI, EDUCATION AND MEANS OF COMMUNICATION

Article - 26

Measures shall be taken to ensure that members of the peoples concerned have the opportunity to acquire education at all levels on at least an equal level footing with the rest of the national community.

Article - 27

1. Education programmes and services for the peoples concerned shall be developed and implimented in cooperation with them to

address their special needs, and shall incorporate their histories, their knowledge and technologies, their value systems and their future social, economic and cultural aspirations.

2. The competent authority shall ensure the training of members of these peoples and their involvement in the formulation and implementation of educational programmes, with a view to the progressive transfer of responsibility for the conduct of these programmes to these peoples as appropriate.

3. In addition, governments shall recognize the right of these peoples to establish their own educational institutions and facilities, provided that such institutions meet minimum standards established by the competent authority in consultation with these peoples. Appropriate resources shall be provided for this purpose.

Article - 28
1. Children belonging to the peoples concerned shall, wherever practicable, be taught to read and write in their own indigenous language or in language most commonly used by the group to which they belong. When this is not practicable, the competent authorities shall undertake consultations with these peoples with a view to the adoption of measures to achieve this objective.
2. Adequate measures shall be taken to ensure that these peoples have the opportunity to attain fluency in the national language or in one of the official languages of the country.
3. Measures shall be taken to preserve and promote the development and practice of the indigenous language of the peoples concerned.

Article - 29
The imparting of general knowledge of general knowledge and skills that will help children belonging to the peoples concerned .to participate fully and on an equal footing in their own community and in the national community shall be an aim of education for these peoples.

Article - 30
1. Governments shall adopt measures appropriate to the, traditions and cultures of the peoples concerned, to make known to them their rights and duties, especially in regard to labour, economic opportunities, education and health matters, social welfare and their rights deriving from this Convention.
2. If necessary, this shall be done by means of written translations and through the use of Mass Communications in the languages of these peoples.

Article - 31
Educational measure shall be taken among all sections of the national community, and particularly among those that are most direct contact with the peoples concerned, with the object of eliminating

prejudices that they may harbour in respect of these peoples. To this, efforts shall be made to ensure that history text books and other educational materials provide a fair, accurate and informative portrayal of the societies and cultures of these peoples.

PART VII, CONTACTS AND COOPERATION ACROSS BORDERS

Article - 32

Governments shall take appropriate measures, including by means

of international agreements to facilitate contacts and cooperation between indigenous and tribal peoples across borders, including activities in the economic, social, cultural and environmental fields.

PART VIII ADMINISTRATION

Article 33

1. The Governmental authority responsible for matters covered in this Convention shall ensure that agencies or other appropriate mechanisms exist to administrator the programme affecting the peoples concerned and shall ensure that they have the means necessary for the proper fulfillment of the functions assigned to them.

2. These programmes shall include :

(a) the planning, coordination, execution and evaluation in cooperation with the peoples concerned, of the measures provided for in this convention;

(b) the proposing of legislative and other measures to the competent authorities and supervision of the application of the measures taken, in cooperation with the peoples concerned.

PART IX. GENERAL PROVISIONS

Article-34

he nature and scope of the measures to be taken to give effect to this Convention shall be determined in characteristic of each country.

Article-35

The application of the provisions of this Convention shall not Adversely affect rights and benefits of the peoples concerned pursuant to other Conventions and Recommendations, international instruments, treaties, or national laws, awards, custom or agreements.

PART X. FINAL PROVISIONS

Article-36
This Convention revises the Indigenous and Tribal Population Convention, 1957.

Article-37
The formal ratifications of this Convention shall be communicated to the Director General of the International Labour office for registration.

Article-38
1. This Convention shall be binding only upon those Members of the International labour Organization whose ratifications have been registered with the Director General.

2. It shall come into force twelve months after the date on which the ratifications of two Members have been registered with the Director General.

3. Thereafter, this Convention shall come into force for any Member twelve months after the date on which its ratification has been registered.

Article 39
1. A Member which has ratified this Convention may denounce it after the expiration of ten years from the date on which the Convention first comes into force by an act communicated to the Director General of the International Labour Office for registration. Such denunciation shall not take effect until one year after the date on which it is registered.

2. Each Member which has ratified this convention and which does not, within the year following the expiration of the period of ten years mentioned in the preceding paragraph, exercise the right of denunciation provided for in this Article, will be bound for another period of ten years and, thereafter, may denounce this Convention at the expiration of each period for in this Article.

Article 40
1. The Director General of the International Labour Office shall notify all Members of the International Labour Organisation of the registration of the registration of all ratifications and denunciations communicated to him by the Members of the Organisation.

2. When notifying the Members of the Organisation of the registration of the second ratification communicated to him, the Director General shall draw the attention of the Members of the Organisation to the date upon which the Convention will come into force.

Article 41

The Director General of the International Labour Office shall communicate to the Secretary General of the United Nations for registration in accordance with Article 102 of the Charter of the United Nations full particulars of all ratifications and acts of denunciations of the preceding Articles.

Article 42

At such times as it may consider necessary the Covering Body of the International Labour Office shall of this Conventions and shall examine the desirability of its revision in whole or in part.

Article 43

1. Should the Conference adopt a new Convention revising this Convention in whole or in part, then, unless the new Convention otherwise provides

(a) The ratification by a Member of the new revising Convention shall ipso jurne involve the immediate denunciation of this Convention 1, notwithstanding the provisions of Article 30 above, if and when the new revising Convention shall have come into force;

(b) as from the date when this the new revising convention comes into force this convention shall cease to be open to rectification by the members.

2.   This convention shall in any case remain in force in its actual form and content for those members which have retained it but have not ratified the revising Convention.

Article 44

The English and French Versions of the text of this Convention are equally authoritative.

The text of the convention as here presented is a true copy of the text authenticated by signatures of the president of the International Labour Conference and of the Director General of International Labour office.

Certified true and complete copy
Sd/-

For the Director-General of the International Labour Office:Francis Maupain Legal Adviser of the International Labour office

Appendix - IX
National Forest Policy, 1988

1. PREAMBLE:-
1.1 In Resolution No. 13/52/F, dated the 12th May, 1952, the Government of India in the erstwhile Ministry of Food & Agriculture enunciated a Forest Policy to be followed in the management of State Forests in the country. However, over the years, forests in the country have suffered serious depletion. This is attributable to relentless pressures arising from over-increasing demand for fuelwood, fodder and timber; inadequacy of protection measures, diversion of forest lands to non-forest uses without ensuring compensatory afforestation and essential environmental safeguards, and the tendency to look upon forests as revenue earning resource. The need to review the situation and to evolve, for the future, a new strategy of forest conservation has become imperative. conservation includes preservation, maintenance, sustainable utilisation, restoration, and enhancement of the natural environment. It has thus become necessary to review and revise the National Forest Policy.

2 BASIC OBJECTIVES:-
2.1 The basic objectives that should govern the National Forest Policy are the following:-

- Maintenance of environmental stability through preservation and, where necessary, restoration of the ecological balance that has been adversely disturbed by serious depletion of the forests of the country.

- Conserving the natural heritage of the country by preserving the remaining natural forests with the vest variety of flora and fauna, which represent the biological diversity and genetic resources of the country.

- Checking soil erosion and denudation in the catchment areas of rivers, lakes, reservoirs in the interest of soil and water conservation, for mitigating floods and droughts and for the restardation of siltation of reservoirs.

- Checking the extension of sand-dunes in the desert areas of Rajasthan and along the coastal tracts.

- Increasing substantially the forest/tree cover in the country through massive afforestation and social forestry programmes, especially on all denuded, degraded and unproductive lands.

- Meeting the requirements of fuelwood, fodder, minor forest produce and small timber of the rural and tribal populations.

- Increasing the productivity of forests to meet essential national needs.
- Encouraging efficient utilisation of forest produce and maximising substitution of wood.
- Creating a massive people's movement with the involvement of women, for achieving these objectives and to minimise pressure on existing forests.

2.2. The principal aim of Forest Policy must be to ensure environmental stability and maintenance of ecological balance including atmospheric equilibrium which are vital for sustenance of all lifeforms, human, animal, and plant. The derivation of direct economic benefit must subordinated to this principal aim.

3. <u>ESSENTIALS OF FOREST MANAGEMENT</u>:-

3.1 Existing forests and forest lands should be fully protected aand their productivity improved. Forest and vegetal cover should be increased rapidly on hill slopes, in catchment areas of rivers, lakes and reservoirs and ocean shores and on semi-arid and desert tracts.

3.2 Diversion of good and productive agricultural lands to forestry should be discouraged in view of the need for increased food production.

3.3 For the conservation of total biological diversity, the network of national parks, sanctuaries, biosphere reserves and other protected areas should be strengthened and extended adequately.

3.4 Provision of sufficient fodder, fuel and pasture, specially in areas adjoining forest, is necessary in order to prevent depletion of forests beyond the sustainable limit. Since fuelwood continues to be the predominant source of energy in rural areas, the programme of afforestation should be intensified with special emphasis on augmenting fuelwood production to meet the requirement of the rural people.

3.5 Minor forest produce provides sustenance to tribal population and to other communities residing in and around the forests. Such produce should be protected, improved and their production enhances with due regard to generation of employment and income.

4. <u>STRATEGY</u>:-

4.1 <u>**Area Under Forests**</u>:- The national goal should be to have a minimum of one-third of the total land area of the country under forest or tree cover. In the hills and in mountainous regions, the aim should be to maintain two-third of the area under such cover in order to prevent erosion and land degradation and to ensure the stability of the fragile eco system.

4.2   Afforestation, Social Forestry & Farm Forestry:-

4.2.1   A massive need-based and timebound programme of afforestation and tree planting, with particular emphasis of fuelwood and fodder development, on all degraded and denuded lands in the country, whether forest or non-forest land, is a national imperative.

4.2.2   It is necessary to encourage the planting of tree alongside of roads, railway lines, rivers and streams and canals, and on other unutilised lands under State/Corporate, institutional of private ownership. Green bells should be raised in urban/industrial areas as well as in arid tracts. Such a programme will help to check erosion and desertification as well as improve the microclimate.

4.2.3   Village and community lands, including those on foresheres and environs of tanks, not required for other development of tree crops and fodder resources. Technical assistance and other inputs necessary for initiating such programmes should be provided by the Government. The revenues generated through such programmes should belong to the panchayats where the lands are vested in them; in all other cases, such revenues should be shared with the local communities in order to provide an incentive to them. The vesting, in individuals, particularly from the weaker sections (such as landless labour, small and marginal farmers, castes, tribals, women) of certain ownership rights over trees, could be considered, subject to appropriate regulations; beneficiaries would be entitled to usufruct and would in turn be responsible for their security and maintenance.

4.2.4.   Land laws should be so modified wherever necessary so as to facilitate and motivate individuals and institutions to undertake tree-farming and grow fodder plants, grasses and legumes on their own land. Wherever possible, degraded lands should be made available for this purpose either on lease or on the basis of a tree-patta scheme. Such leasing of the land should be subject to the land grant rules and land ceiling laws. Steps necessary to encourage them to do so must be taken. Appropriate regulations should govern the felling of trees on private holding.

4.3.   Management of State Forests:-

4.3.1   Schemes and projects which interfere with forests that clothe steep slopes, catchments of rivers, lakes and reservoirs, geologically unstable terrain and such other ecologically sensitive areas should be severely restricted. Tropical rain/moist forests, particularly in areas like Arunachal Pradesh, Kerala, Andaman & Nicobar Islands, should be totally safeguarded.

4.3.2. No forest should be permitted to be worked without the Government having approved the management plan, which should be in a prescribed format and in keeping with the National Forest Policy. The Central Government should issue necessary guidelines to the State Governments in this regard and monitor compliance.

4.3.3. In order to meet the growing needs for essential goods and services which the forests provide, it is necessary to enhance forest cover and productivity of the forests through the application of scientific and technical inputs. Production forestry programmes, while aiming an enhancing the forest cover in the country, and meeting rational needs, should also be oriented to narrowing, by the turn of the century, the increasing gap between demand and supply of fuelwood. No such programme, however, should entail clear-felling of adequately stocked natural forests. Nor should exotic species be introduced, through public or private sources, unless long-term scientific trials undertaken by specialists in ecology, forestry and agriculture have established that they are suitable and have no adverse impact on native vegetation and environment.

4.3.4. **Rights and Concessions:-**

4.3.4.1. The rights and concessions, including grazing, should always remain related to the carrying capacity of forests. The capacity itself should be optimised by increased investment, silvicultural research and development of the area. Stall-feeding of cattle should be encouraged. The requirements of the community, which cannot be met by the rights and concessions so determined should be met by development of social forestry outside the reserved forests.

4.3.4.2 **The holders of customary rights and concessions in forest areas should be motivated to identify themselves with the protection and development of forests from which they derive benefits. The rights and concessions from forests should primarily be for the bonafide use of the communities living within and around forest areas, specially the tribals.**

4.3.4.3. **The life of tribals and other poor living within and near forests revolves around forests. The rights and concessions enjoyed by them should be fully protected. Their** domestic requirements of fuelwood, fodder, minor forest produce and construction timber should be the first charge on forest produce. These and substitute materials should be made available through conveniently located depots at reasonable prices. the area, which such consideration should cover, would be determined by the carrying capacity of the forest.

4.3.5.   Wood is in short supply. The long-term solution for meeting the existing gap lies in increasing the productivity of forests, but to relieve the existing pressure on forests for the demands of railway sleepers, furniture and paneling, mine-pit props, paper and paper board etc. substitution of wood needs to be taken recourse to. Similarly, on the front of domestic energy, fuelwood needs to be substituted as far as practicable with alternate sources like bio-gas, LPG and solar energy. Fuel-efficient "Chulhas" as a measure of conservation of fuelwood need to be popularised in rural areas.

4.4   Diversion of Forest Lands for Non-forest Purposes:-

4.4.1.   Forest land or land with tree cover should not be treated merely as a resource readily available to be utilised for various projects and programmes, but as a national asset which requires to be properly safeguarded for providing sustained benefits to the entire community. Diversion of forest land for any non-forest purpose should be subject to the most careful examinations by specialists from the standpoint of social and environmental costs and benefits. Construction of dams and reservoirs, mining and industrial development and expansion of agriculture should be consistent with the needs for conservation of trees and forests. Projects which involve such diversion should at least provide in their investment budget, funds for regradation/ compensatory afforestation.

4.4.2.   Beneficiaries who are allowed mining and quarrying in forest land and in land covered by trees should be required to repair and re-vegetate the area in accordance with established forestry practices. No mining lease should be granted to any party, private or public, without a proper mine management plan appraised from the environmental angle and enforced by adequate machinery.

4.5.   Wildlife Conservation:-

Forest Management should take special care of the needs of wildlife conservation, and forest management plans should include prescriptions for this purpose. It is specially essential to provide for "corridors" linking the protected areas in order to maintain genetic continuity between artificially separated sub-sections of migrant wildlife.

4.6. **Tribal People and Forests:- Having regard to the symbiotic relationship between the tribal people and forests, a primary task of all agencies responsible for forest management, including the forest development, corporations should be to associate the tribal people closely in the protection, regeneration and development of forests as well as to provide gainful employment to people living in an around the forest. While safeguarding the customary rights and interests of such people, forestry programmes should pay special attention to the following:-**

• One of the major causes for degradation of forest is illegal cutting and removal by contractors and their labour. In order to put an end to this practice, contractors should be replaced by institutions such as tribal co-operatives; labour co-operatives, government corporations, etc. as early as possible.

• Protection, regeneration and optimum collection of minor forest produce along with institutional arrangements for the marketing of such produce.

• Development of forest villages on par with revenue villages.

• Family oriented schemes for improving the status of the tribal beneficiaries; and

• Undertaking integrated area development programmes to meet the needs of the tribal economy in and around the forest areas, including the provision of alternative sources of domestic energy on a subsidised basis, to reduce pressure on the existing forest areas.

4.7. **Shifting Cultivation**:- Shifting cultivation is affecting the environment and productivity of land adversely. Alternative avenues of income, suitably harmonised with the right land use practices, should be devised to discourage shifting cultivation. Efforts should be made to contain such cultivation within the area already effected, by propagating improved agricultural practices. Area already damaged by such cultivation should be rehabilitated through social forestry and energy plantations.

4.8   **Damage to Forests from Encroachments, Fires and Grazing**:-

4.8.1.   Encroachment on forest land has been on the increase. This trend has to be arrested and effective action taken to prevent its continuance. There should be no regulation of existing encroachments.

4.8.2.   The incidence of forest fires in the country is high. Standing trees and fodder are destroyed on a large scale and natural regeneration annihilated by such fires. Special precautions should be taken during the fire season. Improved and modern management practices should be adopted to deal with forest fires.

4.8.3.   Grazing in forest areas should be regulated with the involvement of the community. Special conservation areas, young plantations and regeneration areas should be fully protected. Grazing and browsing in forest areas need to be controlled. Adequate grazing fees should be levied to discourage people in forest areas from maintaining large herds of non-essential livestock.

4.9.   **Forest based Industries:-** The main considerations governing the establishment of forest-base industries and supply of raw material to them should be as follows:

- As far as possible, a forest based industry should raise the raw material needed for meeting its own requirements, preferably by establishment of a direct relationship between the factory and the individuals who can grow the raw material by supporting the individuals with inputs including credit, constant technical advice and finally harvesting and transport services.

- No forest based enterprise, except that at the village or cottage level, should be permitted in the future unless it has been first cleared a careful scrutiny with regard to assured availability of raw material. In any case, the fuel, fodder and timber requirements of the local population should not be sacrificed for this.

- Forest based industries must not only provide employment to local people on priority but also involve them fully in raising trees and raw material.

- Natural forests serve as a gene pool resource and help to maintain ecological balance. Such forests will not, therefore, be made available to industries for undertaking plantation and for any other activities.

- Farmers, particularly small and marginal farmers, would be encouraged to grow, on marginal/degraded lands available with them, wood species required for industries. These may also be grown up along with fuel and fodder species on community lands not required for pasture purposes, and by Forest department/corporations on degraded forests, not end marked for natural regeneration.

- The practice of supply of forest produce to industry at concessional prices should cease. Industry should be encouraged to use alternative raw materials. Import of wood and wood products should be liberalised.

The above considerations will, however, be subject to the current policy relating to land ceiling and land laws.

4.10. **Forest Extension:-** Forest conservation programme can not succeed without the willing support and co-operation of the people. It is essential, therefore, to inculcate in the people, a direct interest in forests, their development and conservation, and to make them conscious of the value of trees, wildlife and nature in general. This can be achieved through the involvement of educational institutional right from the primary stage. Farmers and interested people should be provided opportunities through institutions like Krishi Vigyan Kendras, Trainers' Training Centres to learn agrisilvicultural and silvicultural techniques to ensure optimum use of their land and water resources. Short term extension courses and lectures should be organised in order to educate farmers. For this purpose, it is essential that suitable programmes are propagated through mass media, audio visual aids and the extension machinery.

4.11. **Forest Education**:- Forestry should be recognised both as scientific discipline as well as a profession. Agriculture universities and institutions dedicated to the development of forestry education should formulate curricula and courses for imparting academic education and promoting post graduate research and professional excellence, keeping in view the man power needs of the country. Academic and professional qualifications in forestry should be kept in view for recruitment to the Indian Forest Service and the State Forest Service. Specialised and orientation courses for developing better management skills by in service training need to be encouraged, taking into account the latest development in forestry and related disciplines.

4.12. **Forestry Research**:- With increasing recognition of the importance of forests for environmental health, energy and employment, emphasis must be laid on scientific forestry research, necessitating adequate strengthening of the research base as well as new priorities for action. Some broad priority areas of research and development needing special attention are:-

I. Increasing the productivity of wood and other forest produce per unit of area per unit time by the application of modern scientific and technological methods.

II. Re vegetation of barren/marginal/waste/mined lands and watershed areas.

III. Effective conservation and management of existing forest resources (mainly natural forest eco-system).

IV. Research related to social forestry for rural/tribal development.

V. Development of substitutes to replace wood and wood products.

VI. Research related to wildlife and management of national parks and sanctuaries.

4.13. **Personnel Management**:- Government policies in personnel management for professional foresters and forest scientists should aim at enhancing their professional competence and status and attracting and retaining qualified and motivated personnel, keeping in view particularly the arduous nature of duties they have to perform, often in remote and inhospitable place.

4.14. **Forest Survey and Data Base**:- Inadequacy of data regarding forest resources is a matter of concern because this creates a false sense of complacency. Priority needs to be accorded to completing the survey of forest resources in the country on scientific lines and to updating information. For this purpose, periodical collection, collation and publication of reliable data on relevant aspects of forest management needs to be improved with recourse to modern technology and equipment.

4.15. **Legal Support & Infrastructure Development**:- Appropriate legislation should be undertaken, supported by adequate infrastructure, at the Centre and State levels in order to implement the Policy effectively.

4.16. **Financial Support for Forestry:-** The objectives of this revised policy cannot be achieved without the investment of financial and other resources on a substantial scale. Such investment is indeed fully justified considering the contribution of forests in maintaining essential ecological processes and life-support systems and in preserving genetic diversity. Forests should not be looked upon as a source of revenue. Forests are a renewable natural resource. They are a national asset to be protected and enhanced for the well-being of the people and the Nation.

(K.P.Geethakrishnan)
Secretary to the Government of India